LOCATION
Bibliography

AUTHOR
KINGSFORD

ACCESSION NUMBER
76006

HOUSE OF COMMONS LIBRARY

TO BE
DISPOSED
BY
AUTHORITY

*The Publishers Association
1896–1946*

The Publishers Association
1896-1946
With an Epilogue

R. J. L. KINGSFORD, C.B.E.

Fellow of Clare College, Cambridge;
formerly Secretary to the Syndics of the Cambridge University Press;
President of The Publishers Association, 1943–5

CAMBRIDGE
AT THE UNIVERSITY PRESS · 1970

Published by the Syndics of the Cambridge University Press
Bentley House, 200 Euston Road, London N.W.1
American Branch: 32 East 57th Street, New York, N.Y.10022

© Cambridge University Press 1970

Library of Congress Catalogue Card Number: 74-101445

Standard Book Number: 521 07756 7

Printed in Great Britain
at the University Printing House, Cambridge
(Brooke Crutchley, University Printer)

Contents

Preface *page* ix

Introduction: the 1852 verdict 1

1 1895–1900: the founding of the Association and the Net Book Agreement 5
Macmillan and Marshall, 5; *The spread of net prices*, 7; *Regulation rejected*, 8; *The Association formed*, 8; *The first year*, 9; *Imperial copyright*, 10; *Abridgements and pirates*, 10; *Model agreements*, 11; *Bibliographical practice*, 12; *Obstacles to net prices*, 13; *A trial scheme*, 15; *The Net Book Agreement*, 16; *The International Congress of Publishers*, 17

2 1901–1908: through the Book War 18
Copyright, 18; *The International Congress Office*, 20; *The Education Act of 1902*, 20; *The Net Book Agreement in operation*, 22; *Challenged by 'The Times'*, 22; *The Book War*, 26; *Criticism of the Association*, 35

3 1908–1914: copyright and novel prices 37
New copyright legislation, 37; *The 1911 Act*, 40; *The International Congress*, 42; *The operation of the Net Book Agreement*, 43; *The circulating libraries*, 44; *The sevenpennies*, 46; *Tauchnitz editions*, 48; *The Educational Books Committee*, 49; *First contact with Trade Unions*, 50; *The Publishers' Circle*, 50

4 1914–1919: World War I 51
Leipzig Exhibition, 51; *World War I*, 52; *Paper supplies and prices*, 52; *Shortage of other materials*, 57; *Depleted staffs*, 57; *Rising costs and prices*, 58; *Imports from U.S.A.*, 60; *Copyright in German books*, 61; *National Book Fortnights and Christmas catalogues*, 62; *Export clearing house*, 63; *Threatened legislation*, 66; *Printers and binders: warehousing charges*, 67; *Trade Unions*, 69; *Constitution: Council and Groups*, 69; *Authors and libel*, 71; *End of an era*, 72

5 1919–1927: reconstruction and strikes 73
Council under criticism, 73; *Council elections*, 74; *Exports: opportunities and obstacles*, 75; *Florence Fairs*, 76; *Imports from Germany*, 77; *Revival of the International Congress*, 78; *Customs and copyright*, 79; *Registration at Stationers' Hall*, 82; *Copyright libraries: limited editions and importations*, 83; *Trinity College, Dublin*, 84; *Broadcasting*, 84; *Post-war costs*, 87; *Strikes*, 88; *The Groups*, 90; *The three founding fathers*, 92

Contents

6 1924–1930: new aims and opportunities   page 94
The Society of Bookmen, 94; The National Book Council, 95; Booksellers and the Public Libraries, 95; The Co-operative Societies, 97; The visit to Germany and Holland, 98; 'The Joint Committee', 99; The standing Joint Advisory Committee, 102; Interpretation of the Net Book Agreement, 103; The Library Agreement, 104; The official trade paper and trade directory, 105; Commonwealth booksellers, 109; Australian terms and prices, 109; Australian economic difficulties, 111

7 1931–1939 (1): the Great Depression, markets and rights   113
Devaluation and import duties, 114; Exchange control and restriction of credit, 115; The British Council, 117; Retention of markets, 118; Cheap Continental editions, 119; Dominion markets, 120; Local piracy, 121; The Indian Group, 122; Copyright in typography, 123; Broadcasting and other subsidiary rights, 124; Guide to royalty agreements, 126; Terminable licences, 127; Libel and obscene libel, 128; The International Congress, 130; The Weimar resolution and freedom of publication, 131

8 1931–1939 (2): book clubs, Book Tokens, book weeks   133
Book clubs, 134; Book Tokens, 138; Gift coupons, 145; Coupon advertising, 147; Twopenny libraries, 149; 'The Joint Committee' revived, 150; The Joint Advisory Committee at work: 'other traders', book agents, quantity discount scheme, 151; The Net Book Agreement and the 'Co-ops', 152; Book weeks and exhibitions, 153; Re-organization and new Rules, 154; The Groups, new and old, 155; The Book Manufacturers' Association, 156

9 1939–1946 (1): World War II: the book front   159
'THE PHONEY WAR': War risks insurance, 160; Paper, 161; Labour and other restrictions, 163; Faber's conclusions, 164; The Association's set-up, 164
1940–1941: Paper, 165; The Export Group, 166; The Blitz, 167; Simpkin's, 167; The purchase tax battle, 168; For the Forces, 169
1941–1942: Labour, 170; Paper economy, 171; The Moberly Pool, 172; Cloth and boards and salvage, 174
1942–1943: Board of Trade as sponsors, 176; Economy standards, 176; Metal, 177; Men and women, 177
1943–1944: Paper, 178; 'Free' paper, 179; Discriminatory pressure, 180; The small publisher, 180; Labour, 181; The Forces, 181; Europe, 182
1944–1945: The parliamentary deputation, 182; Quota and pool, 183; 'Free' paper again, 184; Electoral registers, 184
1945–1946: Paper, 184; Economy standards ending, 185; Labour, 185

10 1939–1946 (2): World War II: the export markets, trade relations at home   187
THE EXPORT MARKETS: Book Export Scheme, 187; South America, 188; The liberation of Europe, 188; American competition, 189; Canada, 190; Australia, 191; Hitchcock's mission, 191; The Association's mission, 193; In New York, 193; In Canada, 194; The mission's report, 194; The Canada Committee, 195; The U.S.A. Committee, 195; Agreement reached, 196; Export research, 197

TRADE RELATIONS AT HOME: *The Joint Advisory Committee*, 198; *Commercial lending libraries*, 198; *Partial remaindering*, 199; *National Book League*, 200; *Society of Authors*, 201; *The coming of the microfilm*, 203; *International copyright*, 203; *Printers and binders*, 204; *Binding research*, 204; *Domestic affairs*, 205; *The Association's jubilee*, 206

Epilogue: the 1962 verdict   *page* 207

Appendices
  1 *The first Rules of the Association*   213
  2 *The members in 1896 and the first Council*   216
  3 *The Officers and Secretaries of the Association, 1896–1962*   218

Bibliography   219

Index   221

Preface

This is a history of the first 50 years of the Association established by British book-publishers for the regulation of the common policies and practices of their trade. The picture is framed by two contrary verdicts on resale price maintenance, of which the first in 1852 surely postponed by some 40 years the founding of the Association and the second in 1962 vindicated the Net Book Agreement, which the Association had promoted as its first official act.

I have had much helpful criticism: from John Brown, who indeed encouraged me during his Presidency of the Association to attempt this incursion from publishing into authorship; from J. D. Newth; from F. D. Sanders and R. E. Barker, the past and present Secretaries of the Association; and from my son, A. L. Kingsford, who has also read the proofs. To all of them and to Miss Alice Cottam, whose typing of my MS. greatly eased my task, I am truly grateful. Sir Stanley Unwin and G. Wren Howard also read the chapters covering the years in which they were most active in the leadership of the Association, but them, alas, I can no longer thank. I am glad also to acknowledge useful information from Miss Livia Gollancz, John Baker, Harold Raymond, and Henry Schollick.

The archives of the Association have been made available to me without reservation.

Finally, I am grateful to the Syndics of the Cambridge University Press for undertaking the publication. It has been a pleasure once more to work, although in a different capacity, with the staff of the Press.

R. J. L. K.

1 *January* 1970

Note

The publishing firm to which an individual belongs, unless self-evident, is named in the text upon his first appearance and in the index.

Introduction
The 1852 verdict

> Such regulations [for price maintenance] seem *prima facie* to be indefensible, and contrary to the freedom which ought to prevail in commercial transactions.
>
> Lord Justice Campbell, 1852

In the long history of the production and sale of books the separation of publishing and bookselling as distinct functions is a recent growth. Joint publication by share-holding booksellers continued into the nineteenth century, and in the absence of one responsible publisher it was indeed the name and address of the printer which were required by law on the title-page of a book. Copyrights were acquired jointly at auction sales and as many as fifty participants would share the expenses of the publication and receive copies of the book in proportion to their investment. In the eighteenth century the publishers, or wholesalers, would normally issue a book in unbound sheets and even when, in the first quarter of the nineteenth, binding in cheap boards became possible, uniformity of style would not necessarily be maintained and the fixing and maintenance of a retail price certainly were not. The 'discount question'—meaning price-cutting or free trade or literary protection, as it was seen through varying eyes—had come and had come to stay.

Joint publication inevitably produced problems and disputes between the partners, requiring an organized solution. Temporary committees of London booksellers were set up in 1808 to consider a new Copyright Bill then before the House of Commons, and in 1828 to make recommendations for the regulation in management and finance of share-holding enterprises. A period of trade depression in 1825–6, leading to a financial crisis which some publishing booksellers did not survive, and the successful co-operation in the Committee of 1828 led in the following year to a meeting of a Committee of London booksellers, held at the Chapter Coffee House, at which it was resolved that 'the practice of underselling having grown to an alarming extent, both by advertising...and also by ticketing of books, by which means great loss had been occasioned to respectable booksellers, and publishers, and serious injury done to the respectability

2 Introduction: the 1852 verdict

of the trade' measures should be taken to check such practice and to increase the respectability of the trade. It was then agreed that no reduction of the published price was permissible except one of 10% for cash payment and an allowance of not more than 15% to book clubs and reading and literary societies; that any persistent infringer should have his name erased from the list of booksellers; and that the committee should become a permanent one, to give effect to these regulations. The growth of advertising, or 'puffing' as it came to be called by those who regarded it as an unrespectable practice, was beginning to make book-buyers price-conscious.

The committee set up in 1829 continued to be effective and, in enforcement of the regulations, to publish lists of undersellers. Although reaction by rebels strengthened, it was not until 1852 that the internal dissension of the book trade became a matter of widespread public interest; and then, fanned by *The Times*, a great controversy broke out.

Apart from the fact that in 1850 the booksellers in general meeting had reiterated the regulations of 1829 and that price maintenance could be presented, by doubtful analogy, in the context of the corn laws and of free traders versus literary protectionists (with booksellers likened to farmers and authors to cotton planters) it is difficult to see why it should have seemed a greater public menace then than at any time during the previous twenty years. In March 1852 *The Times* thundered in a leading article against 'this anomalous interference with the free course of competition and the natural operations of trade'; letters followed in its correspondence columns, signed and anonymous (including one from 'Two and Two' who, as others have done since, got his figures wrong and confused the bookseller's total allowance with his profit); and Gladstone spoke on the matter in the House of Commons. In the meantime the protagonists, the Booksellers' Committee and the rebels, were referring their dispute to arbitration and agreeing to accept the verdict, whichever way it should go. In January John Chapman,[1] a bookseller who specialized in the import of American books, had launched an attack on the Booksellers' Association, challenging in particular its jurisdiction over foreign books. The reply of William Longman[2] and John Murray[3] on behalf of the Committee was

---

[1] This is the John Chapman with whom George Eliot lodged; see Haight, *George Eliot and John Chapman* (New Haven, 1940).

[2] This is the Longman whom Trollope apostrophized in *Barchester Towers* (1857): 'a difficulty begins to make itself manifest in the necessity of disposing of all our friends in the small remainder of this volume. Oh, that Mr Longman would allow me a fourth! It should transcend the other three.'

[3] John Murray III, the grandson of the founder of the firm and the son of John Murray II, who had published Byron and Scott and had founded *The Quarterly Review*.

Introduction: the 1852 verdict

to invite Lord Campbell, later Lord Chancellor, with other eminent men to be chosen by him from names suggested, to arbitrate; and as the Committee's decision to abide by the decision and to resign if it were unfavourable came to be regarded as binding upon the whole Association, the Association was in the event fighting for its life. Lord Campbell selected George Grote, the historian, and H. H. Milman, Dean of St Paul's, as his colleagues, and beginning on 14 April the arbitrators received oral evidence from the Committee and a written record from the dissentients which included letters from such notable figures as Carlyle, Darwin, Dickens, Mill and Tennyson. The issues were argued in terms of commerce, of free trade or monopoly, with analogies from Mr Cobden's manufacture of muslins and Mr Bright's new line in carpets, and of the irreparable financial consequences to publishing and bookselling if the undersellers were allowed to continue without restraint. With hindsight we may now be surprised that it was assumed that the maintenance of a fixed price would inevitably cause the book-buyer to pay more and that no one argued that if the bookseller were prevented from giving away part of his trade allowance, the publisher could reduce it and so reduce also his fixed price. The arbitrators gave their decision on 19 May, and one may be permitted to wonder whether they had discussed the propriety of Gladstone's speech in the House of Commons seven days earlier, in which he praised the enterprise and energy of the undersellers and said that he could not much doubt what the verdict of the arbitrators would be. The decision was unequivocal: the regulations were harmful and vexatious and inconsistent with principles of free trade, and authors, the public and even booksellers would benefit from their abandonment. Platitudinous and shallow though Lord Campbell's verdict now seems to have been, it was effective. The Booksellers' Committee resigned, and although a new committee was set up to consider continuing co-operation within the trade, it could not recommend any alternative to unlimited competition and the Association ceased to exist.[1]

In spite of, or because of, his victory Chapman himself was insolvent by 1854, but underselling continued. Efforts were made unsuccessfully in 1859–60 to restart a Booksellers' Association; and the London Booksellers' Society, founded in 1890, failed to persuade the leaders of the now

---

[1] Chapman printed *A Report of the Proceedings of a meeting (consisting chiefly of authors) held May 4th, at the house of Mr. John Chapman, 142 Strand, for the purpose of hastening the removal of the Trade Restrictions on the Commerce of Literature* (London, 1852). In 1906 in a pamphlet entitled *Publishers and the Public*, *The Times* reprinted the leading articles, reports of speeches and correspondence which it had printed in 1852. The story is also told at length in Barnes, *Free Trade in Books* (Oxford, 1964).

separate publishing trade to enforce a system of 'net' published prices. In 1852 the book trade had entered a period of free trade which was to last until the end of the century, and it was to wait for 110 years for a judgment contrary to the verdict of Lord Campbell. While during the latter half of the nineteenth century the prosperity and number of booksellers declined, the publishing trade enjoyed considerable expansion. Between 1801 and 1851 the population of Great Britain had almost doubled; in the seventy years since 1780 wages generally had increased by 50–100%; and in the second quarter of the century the demand for books, presumably from the new commercial classes, produced a decrease in the average price of new works from 16s. to 8s. 4½d. After the Elementary Education Act of 1870, which was to create in due course an entire population able, even if unlikely, to read books, the gradual decline in illiteracy and the immediate demand for school textbooks within a population still rising in number and in income resulted in an increase in the number of new books published annually, from approximately 2,600 in the middle years of the century to 6,044 in 1901. In this large increase religion had no share, but in novels and juvenile books, in history and biography, in economics and economic history, and in poetry and drama there were substantial increases.

The years 1895 and 1896, in which the history of the Publishers Association begins, saw the publication of Hardy's *Jude the Obscure*, Housman's *A Shropshire Lad*, Meredith's *The Amazing Marriage*, Stevenson's *Weir of Hermiston*, and Lewis Carroll's *Symbolic Logic* (accompanied by a folder containing nine counters, four red and five grey); Kipling, Maitland and the Webbs were in their prime as authors and Bernard Shaw was writing his early plays and Conrad his early novels. William Morris, to whose influence the coming renaissance of English printing was to owe so much, died in 1896, and that year brought also the birth of the *Daily Mail* and the repeal of the statute which required power-driven vehicles to be preceded by a man carrying a red flag. Then as now London was the main centre of book publishing. Of the 58 firms who were admitted to membership of the Association within its first year 53 were in London, 4 in Edinburgh and 1 in Bristol; and of the 53 in London 17 were housed within the shadow of St Paul's and 16 in the Covent Garden–Strand neighbourhood. The names of approximately half of the 58 still appeared on the title-pages of the new books of the 1960s.

1

1895-1900
The founding of the Association and the Net Book Agreement

When on 23 January 1895 booksellers from London and all parts of the country met to found the Associated Booksellers of Great Britain and Ireland, they believed that a collapse of retail bookselling was imminent, and the only immediate aims of the Association were to persuade publishers systematically to adopt 'net' prices—fixed retail prices from which no discount could be given—except for such special categories as school books which would be subject to a reduced discount, and to convince them of the need to enforce the system.

MACMILLAN AND MARSHALL

The fixing of net prices for selected books had developed in the hands of one firm during the preceding five years. In 1890 Frederick Macmillan published Alfred Marshall's *Principles of Economics* at a net price, and it has been accepted that its success and Marshall's willing agreement to the fixing of a net price played a decisive part in the establishment of a net price system. That by successfully publishing *Principles of Economics* Macmillan inaugurated the system there can be no doubt, but since the publication in 1965 of Marshall's part in the correspondence[1] it must be conceded that Marshall's acceptance was not accompanied by full understanding of what was meant.

Before 1890 Macmillan had fixed some net prices but always for 'large books selling for several guineas'[2] and not for books expected to have a wide sale at moderate prices; and Macmillan's original proposal in March 1890, which Marshall accepted, was that the price of *Principles of Economics*

---

[1] C. W. Guillebaud, 'The Marshall–Macmillan Correspondence over the Net Book System', *Econ. Journal*, LXXV, no. 299 (1965).

[2] Macmillan to Marshall, 15 April 1890. It should be noted that there were net prices before 1890.

6 1895–1900: the founding and the Net Book Agreement

should be 16s. subject to the usual (variable) discount. In the following month Macmillan wrote[1] that he had become convinced that present discounts of twopence or even threepence in the shilling produced two evils: the inflation of nominal prices, to give room for the discount, and price competition between booksellers, which had so reduced their profit that they held inadequate stocks and gave inadequate service and only remained in business by selling Berlin wool and other fancy goods. He was also convinced in the absence of uniformity of opinion among publishers, booksellers and authors, that a net book system could only be introduced by stages. He then proposed that the price of *Principles of Economics* should be 12s. 6d. net, the trade price to be 10s. 5d. with a 5% discount on settlement, 25 copies not to be charged as 24 or 13 as 12½, as was then generally done. Marshall's book, he wrote, would be an ideal choice for experiment: it was important to choose a good book because if it did not sell, its failure would be attributed to its 'netness' and not to its quality; and it would not 'in any case come in the way of the "cheap-jack" booksellers, who are the only opponents of our scheme'. Macmillan's choice of a book by Marshall was indeed an inspired one, because economists (for example, Mill in the 1852 Book War) had generally opposed fixed prices as a restrictive practice.

Marshall accepted the net price, but before the end of the month of publication he was advocating a net system under which the published price would be the one charged by the bookseller on short credit, a discount of 1d. in the shilling would be given for instant cash, and an addition of 5% per half-year would be charged for long credit.[2] His belief that customers paying cash should be rewarded and those taking long credit should be penalized continued to grow, and he was to return to it in letters to Macmillan in 1897[3] and 1898. Macmillan replied—and it is hard to see how he could have done other—that 'as the competition which had brought retail bookselling to the verge of ruin came about through the pretence of giving discounts for ready money, it would be very dangerous to begin the same system with *net* books' and that 'the only safe plan is to treat all bookselling as if it was a cash business (which for the most part it is) and to make no provision for long credit'.[4] Macmillan's reply and the heat of September 1898 drew from Marshall the

---

[1] *ibid.*
[2] Marshall to Macmillan, 28 July 1890.
[3] 'We [academic economists] are agreed that a trader may fitly borrow from the private person (e.g. via bank): but that when the consumer borrows from his shopkeeper it is economically forcing water to run uphill, and morally harmful.' 3 December 1897.
[4] Macmillan to Marshall, 15 September 1898.

rejoinder: 'I cordially approved the net system for my book when it was suggested to me. But that was because I misunderstood the proposal... The Meteorological Office makes me hope that it will have rained before this reaches you; and if so you may be inclined to forgive me.'[1] In that letter and in subsequent ones in the next month it becomes clear that his main wish is that what he called 'grave scientific books' should, together with school books, be outside the net system so that scholars, with lower incomes than buyers of general literature, should pay less, and because the bookseller's margin should be less on a category of books of which he could do little or nothing to advance the sale. We may think that this was an argument, not against the net book system, but for varying trade margins on different kinds of order—whether for stock or a chance order and whether large or small—which were to become accepted practice within the net book system. 'One thing at least is clear,' concluded Mr Guillebaud from his reprint of the correspondence, 'Marshall's criticisms of the working of the net book system were not based on opposition to the principle of resale price maintenance when applied to books.'

THE SPREAD OF NET PRICES

During 1890 Macmillan published other books at net prices and his lead was quickly followed by other publishers, but when in 1894 the London Booksellers' Society tried to persuade Longman, Murray, Bentley, Black, Smith-Elder, and Blackwood to enforce their prices, they refused to do so; and in a printed circular dated 11 April in that year the Society confined itself to saying that its principle was to limit the discount to 25% and asking booksellers whether, in particular, they wished Warne's edition of Nuttall's Dictionary, published at 3s. 6d., to be sold at less than 2s. 8d. for cash.[2]

From its foundation in the following year the Council of the Associated Booksellers urged booksellers and publishers to support the net book principle and on 10 July a circular was sent to all publishers requesting a formal meeting and proposing that all books published at net prices should be sold at the full price, that prices of 7s. 6d. upwards should be net prices and that the maximum discount allowed on other books should be 25%.

[1] Marshall to Macmillan, 17 September 1898.
[2] London School of Economics: Library of Political and Economic Science, Coll. G.403.

8 1895–1900 : *the founding and the Net Book Agreement*

REGULATION REJECTED

Later in the month a small meeting of publishers was held at John Murray's publishing house, then as now at 50 Albemarle Street, and at their request C. J. Longman wrote on 29 July to the Secretary of the Associated Booksellers, reminding him of Lord Campbell's verdict of 1852 and deducing that any combination of booksellers and publishers originated for the purpose of regulating prices to the public and enforced by coercion or exclusive dealing was altogether unpracticable, and advancing a suggestion, which he had made at the meeting, that fixed retail prices should be abandoned and that publishers should fix wholesale prices and announce them to the public in their advertisements. Perhaps not surprisingly Longman's letter, of which he sent copies to the trade papers,[1] did not commend itself to the booksellers.

THE ASSOCIATION FORMED

Nevertheless, a meeting of publishers and booksellers was held at Anderton's Hotel in the Strand on 10 October at which it was agreed that a meeting of publishers should be summoned to consider the desirability of appointing a Committee to meet the Committee of the Associated Booksellers. The meeting took place on 21 November at Stationers' Hall[2] and Longman was voted into the chair. After a unanimous vote against the appointment of a Committee on the ground that it would be unfair and misleading to the booksellers in encouraging them to expect some concession which it was not possible for publishers to make, it was agreed, again unanimously, that a Committee be appointed to draw up rules for the formation of a Publishers Association, and the following firms were elected to form a committee of nine: Longmans, Green & Co.; Macmillan & Co.; John Murray; Routledge & Sons; William Heinemann; Sampson Low, Marston & Co.; R. Bentley & Son; William Blackwood & Sons; Smith, Elder & Co. After two meetings in December (Longman continuing in the chair and R. B. Marston acting as Secretary) the Committee submitted draft rules to a general meeting held on 23 January 1896 and attended by representatives of forty-nine firms, including some whose eligibility was uncertain or to whom by accident an invitation had not been sent. Rules were agreed[3] with the addition of a wise precaution against the possibility of continuous re-elections of the same persons to

---
[1] *The Publishers' Circular*, 3 August 1895.
[2] which was to be the place of publishers' meetings until its partial destruction by bombing in December 1940.
[3] see App. 1.

the offices of President and Vice-President, and notwithstanding the difficulty of defining a book-publisher. Anyone it was said, not for the last time, could call himself a publisher; and it may be thought that the problem was hardly solved by acceptance of the eligibility of 'individuals who for not less than one year have carried on the work of *bona fide* book publication'. The existing Committee was empowered to accept or reject applications for membership and, when the Association had thus been brought into existence, would then dissolve. The Committee went to work and when the first constitutional meeting was held on 12 March, fifty-five firms had been admitted and three more were added before the end of the year.[1] The meeting was informed that a bank account had been opened and it proceeded to the first elections: C. J. Longman, President; John Murray,[2] Vice-President; Frederick Macmillan, Honorary Treasurer; and ten firms to form the first Council.[3] It was agreed that the Rule which provided that member firms must be represented by partners or directors should be amended to include also managers or secretaries and finally that voting by proxy should not be permissible. The Publishers Association had been founded.

THE FIRST YEAR

The Council met for the first time on 26 March 1896 and found plenty of work requiring its attention. It gratefully accepted an offer from the Stationers' Company of office accommodation in Stationers' Hall and of the part-time services of William Poulten, Beadle of the Company, as a paid Secretary[4] and by the end of its first year it had held thirteen meetings, as against the four prescribed in the Rules, and had appointed six sub-committees. In the inaugural address on 21 April[5] C. J. Longman, to whose initiative and wisdom the infant Association owed a particular debt, called attention to three matters requiring the collective attention of the trade. He outlined the *desiderata* in the law of copyright—compre-

[1] see App. 2. The eligibility of Oxford and Cambridge University Presses may have seemed doubtful. Cambridge, then in partnership with C. J. Clay & Sons, backed both horses and the application from Clay was the one accepted. Oxford made no application until 1898. It may also be noted that at the Annual General Meeting in 1904 G. Gill & Sons moved that an inquiry be made into the right of the University of Cambridge under its charters to carry on business as printers and publishers and as an examining body. The motion was referred to the Council, which after an investigation by the Officers decided that there was no *prima facie* cause of complaint justifying a committee of inquiry.
[2] John Murray IV, later Sir John Murray, K.C.V.O. (died 1928).
[3] see App. 2.
[4] at £1 per week, increased in 1897 to 30s. The office of Honorary Secretary was held by R. B. Marston of Sampson Low, Marston & Co. (see p. 106) until 1904 when he resigned and the office was discontinued.   [5] *P.C.*, 25 April 1896.

hensibility, liberality, universality, and ease of enforcement—and drew attention to a threat from Canada to universality even within the British law. He deplored antagonism, held by some to be natural, between author and publisher and regretted that the differing points of view which were inevitable between seller and buyer were aggravated by lack of clarity in publishers' agreements or even by the absence of any written agreement. Lastly, while expressing to the booksellers his sympathy with any proposal which stood a reasonable chance of mitigating their financial difficulties, he saw no good purpose in reviving recent proposals for a ring to raise the effective retail price and to enforce it.

IMPERIAL COPYRIGHT

Longman's hope of early improvement in Imperial copyright was not fulfilled. Two Bills, to the drafting of which the Association contributed, were produced by the Society of Authors[1] and the Copyright Association and were introduced into the House of Lords by Lord Monkswell and Lord Herschell respectively. Both were referred to a Select Committee, before which the Association gave evidence; a redraft by Lord Thring passed the House of Lords in 1900 and was forwarded to the Colonial Governments for their approval.

In 1896 the Association in concert with the Society of Authors and the Copyright Association represented to Mr Secretary Chamberlain at the Colonial Office the danger of any departure from the existing uniformity which secured for any book first published in the British Empire copyright within all countries signatory to the Berne Convention for forty-two years, or the lifetime of the author and seven years thereafter, whichever were the longer. There was a threat of separate Canadian legislation, but reassurance was received from the Governor-General and satisfactory discussion took place with a representative of the Canadian Authors' Society during the third International Congress of Publishers[2] in 1898; and although again in 1900 a Bill drafted for introduction in the Dominion Parliament gave protection only to books printed in Canada, the legislature agreed to await the expected new Imperial Act.

ABRIDGEMENTS AND PIRATES

Other problems related to copyright were less intractable. In November 1896 the Council agreed that a warning should be sent to the press generally, and to the editors of literary periodicals in particular, against

[1] The Chairman of the Society was then H. Rider Haggard, of *King Solomon's Mines* fame.
[2] see p. 17.

the unauthorized quotation of excessive extracts from copyright books in articles purporting to be reviews. 'It is believed' wrote C. J. Longman[1] 'that articles of this nature satisfy the desire of the public for information and enable them to learn the essential points in a book without buying it.' Allied to this problem was the publication of abridgements of popular novels in such series as *Popular New Novels*, *The Masterpiece Library*, and the *Review of Reviews*, all edited by W. T. Stead. With the backing, and at the expense, of the Association, which took the opinion of Cozens Hardy, Q.C., the abridgement of Mrs Humphry Ward's *Sir George Tressady* in the *Review of Reviews* was made a test case by Messrs Smith, Elder. Stead submitted to a restraining injunction and gave a general undertaking; and this nuisance was brought to an end.

The first action of the Council at its first meeting in March 1896 was to set up a Colonial Piracies Sub-committee with instructions to appoint local agents armed with powers to stop the importation of pirated editions, chiefly of American origin. By February 1898 agents had been appointed in the provinces of Australia, New Zealand, the South African colonies, the Indian presidencies, Ceylon, Singapore and Hong Kong, and powers of attorney had been given to them by most members; and to facilitate economy and clarity of communication, a telegraphic address for the Association was registered in that month (the word chosen, with a surprising lack of appropriateness, being Euphuism)[2] and a series of coded instructions was devised.[3] Legal action was taken in several cases, most notably for the seizure of 101 school books in Calcutta in January 1899. Within its first year the Council was urging upon members the importance of retaining colonial and continental rights and of not surrendering Canadian rights in agreements with American publishers.

MODEL AGREEMENTS

The insistence of the Council upon the need for British publishers to retain colonial rights and the English language rights on the Continent may have been the cause of an early disagreement with the Society of Authors. Following Longman's hope, expressed in his Inaugural Address, for better relations between publisher and author, and for the regularization of agreements, the Council immediately sought from members (and perhaps surprisingly obtained from a body of individualists not yet

---

[1] *Report of the Council for 1896–97*. The Annual Reports, unbroken in series, provide convenient summaries of the principal events of each year.
[2] Subsequently abandoned.
[3] e.g. 'claret' = Do you wish me to proceed and in whose name?

accustomed to co-operation) copies of their standard forms of agreement with authors. Model forms, in skeleton, on royalty, profit-sharing, and commission terms were prepared by a sub-committee and approved as 'suitable models for the trade' by Joseph Walton, Q.C., and in March 1898 their impending circulation to members was announced.[1] A year later in the Annual Report of the Council, the President, John Murray, wrote:

A copy was also officially sent as a matter of courtesy to the Authors' Society, in accordance with the intention expressed by Longman in his presidential address in 1896. Our members are all aware that the drafts were never intended for use in the exact form in which they stand, but were regarded as a collection of clauses adaptable to different cases, drawn up in an approved legal form. It had always been the hope of your Council that, if the authorities of the Authors' Society found anything objectionable in these drafts, they would ask for an explanation, and would probably propose a conference between representatives of the two Societies to discuss, and if necessary to modify, the forms of agreement. The Society, however, took another view of the case, and have, we believe, issued a very severe criticism of the drafts. No copy of this criticism had been communicated to your Council.

The possibility of two-sided discussion was missed.

BIBLIOGRAPHICAL PRACTICE

From what was somewhat cryptically called the Committee on Title-pages came in November 1897 recommendations for the better conduct of publishers in printing—and not suppressing—the bibliographical history of their publications. The terms 'impression', 'edition' and 'reissue' were defined (without the nicety which bibliographers now attach to them); every title-page should bear the date of publication, i.e. the year in which the impression or the reissue was put on the market; a reissue should be described as such; the date at which a book was last revised should be stated; and the bibliographical history should be printed on the back of the title-page so that it would not be detached from it in binding. That publishers needed discipline in these matters is to be

[1] The model forms were printed in 1898 and reprinted in 1904. Provisions in them which may be particularly noted were: that the publisher should have the exclusive right of publication in the English language throughout the world and, with the consent of the author, might sell serial and other rights; that the sale of 13 copies was to be reckoned as 12, or 25 as 24; that the author was to be allowed free corrections in proof up to 25% of the cost of composition; and that the right of publication would revert to the author if at the end of three years from the date of publication the book should have been out of print for six months. The model forms did not suggest the rate of royalty to be paid or, in a profit-sharing agreement, the proportion of the division of profits or the deduction to be made to cover the expenses of publication before the division of the profit.

deduced from the repetition of the recommendations on four occasions before 1905.[1] The committee also effected an improvement in the form in which new publications were listed throughout the year in *The Publishers' Circular* and in the annual 'London Catalogue'.

OBSTACLES TO NET PRICES

After Longman's Inaugural Address in April 1896 the Associated Booksellers continued to press for all prices above 7s. 6d. to be net prices and for the limitation of the discount allowed to the public on non-net prices to 25%, and the Council continued to reject the pressure on the grounds that prices were a matter for individual publishers and that the limitation of the discount would not be enforceable. But at the end of the year, convinced perhaps by figures prepared by Robert Maclehose,[2] of the Glasgow University booksellers of that name, and alarmed by the continued decline in the number of booksellers with real knowledge of their trade,[3] the Council appointed a Committee[4] to discuss with a committee of the Associated Booksellers the question of trade terms and especially the possibility of a reversion to a maximum allowable discount of twopence in the shilling. The Sub-committee on Trade Terms, as it was called, held its first meeting, with a committee of the Associated Booksellers, on 13 January 1897 and it was agreed that the latter should obtain by a circular the opinion of their members on a reversion to twopence in the shilling. It was some months before the replies were sufficiently complete to be representative, but at its meetings in May and June the Council of the Publishers Association agreed to call a special General Meeting and to recommend to members a trial of a scheme under which present trade terms would be given only to booksellers who agreed not to allow more

---

[1] At the Council meeting in December 1902 a list was tabled of seventeen member firms who had recently published books with no dates or with incorrect dates. Disregard of the recommendations came principally from firms selling by door-to-door canvassing.

[2] Maclehose, *The Report of the Society of Authors on the Discount Question. A Criticism* (Glasgow, 1897). Maclehose's published statement showed that, with very few exceptions, retail bookselling of new books with a discount of 25% to the public was an impossibility: that on a turnover of £10,500 there would be a profit of £170 and on £6,000 a loss of £20. If it assumed, as appears to have been accepted at the time, that a bookseller's overhead expenses ran at 15% of his receipts, a 6s. novel bought from the publisher at 4s. and retailed at 4s. 6d. showed him a loss of 2d. and it was only on books which he could buy in quantity, thirteen as twelve, or by a settlement discount for prompt payment that he could turn the loss into a profit.

[3] Maclehose (*op. cit.*): 'It has been stated on good authority that only 200 booksellers who have a real knowledge of their business survive out of 1,200 some twenty years ago, and the number is rapidly decreasing.'

[4] The members were C. J. Longman, F. Macmillan, W. Heinemann, S. Tanner (A. D. Innes & Co.), T. Fisher Unwin.

than twopence in the shilling discount on 'ordinary books' and to sell net books at full price. At the meeting on 1 July the scheme was proposed by Frederick Macmillan, who reported that of 789 booksellers circularized 729 had assented, 12 had declined and 48 had not answered; the scheme was unanimously adopted, and the Council was requested to secure the co-operation of the Society of Authors. It is worth noting that the Association thought it necessary to obtain the co-operation of the Society and the understanding of the public; and considerable publicity was given to the scheme in the press. The issue which the Association hoped that the Society would face was whether the small increase in retail prices would materially reduce the sale of books or whether the greater ability of booksellers to stock and display books, which their better profits would give them, would result in larger sales. The Society, after hearing evidence from booksellers, both for and against, appears to have limited its diagnosis to the effect of the scheme upon the 6s. book, i.e. the novel, and likening books to patent medicines saw no greater reason to improve the position of the booksellers than of the chemists. Although, as the *Pall Mall Gazette* commented in September, the public would have to pay 5s. for its 6s. novels, 'it will have a "bookseller" in the real sense of the word to buy from, instead of the usual combination of stationer and draper'. But the issue, as it seemed to the Society, was not whether there should be control of prices, but how control would work, and in a report bristling with distrust of publishers fear of coercion predominated. The profits of publishers as well as booksellers would be increased; and 'the independence of the author would be seriously compromised by the existence of a close ring of publishers and booksellers who might as easily dictate to him a royalty of 5% as to the bookseller a 2d. discount'.[1] Though it may be doubted whether the Association would have agreed to widen the discount question into a discussion of publishers' profits and advertising expenditure, as was demanded, in the face of the Association's expressed unwillingness to go forward without the co-operation of the Society, the reply was an uncompromising one, and the comment of *The Athenaeum* seems just. 'The Society' it wrote,

may be wise in its resolution—as we do not think it is—but it has certainly been unwise in giving its reasons. The main one is oddly indicative of that belief that the publisher is an hereditary foe which possesses the Society. We are gravely told that if the Publishers Association succeeded in its plan of refusing trade discounts to cheapjacks, it might proceed to dictate to retail booksellers what books they should sell, and thus force authors to publish with members of the

[1] Maclehose, *op. cit.*

Association on any terms those monopolists chose to grant. In view of this supposed future danger from the villain publisher, the country booksellers, on whom all authors but a few popular novelists depend for the distribution of their works, are to be sacrificed.[1]

A TRIAL SCHEME

At a Council meeting called specially in December it was agreed not to proceed with the scheme, but hope was expressed of finding some means of overcoming the obstacles.[2] In April 1898 a new proposal, originated by Robert Maclehose, for the continuance of the threepence in the shilling discount on the 6s. book came forward from the Associated Booksellers, which also referred it to the Society of Authors direct; and when at its May meeting the Council learned that the Society's second report was not unfavourable and so diametrically opposed to its first, it feared a misunderstanding. Accordingly in August the Sub-committee on Trade Terms convened a meeting with representatives of the Associated Booksellers and its allied regional associations and with G. H. Thring, the Secretary of the Society. Thring said that the explicit mention of coercion in the former scheme had been the ground of its rejection, but the new plan did not involve coercion; whereas under the former scheme a bookseller's whole account might be closed, the penalty under the new scheme was refusal to supply a particular book except on stringent terms. So far as the Society was concerned the way ahead was clear, and although the conference saw difficulties in bringing the wholesale booksellers into line and noted that the twelve booksellers who were opposed to the new scheme (as opposed to 700 in favour of it) were the most influential in London, it nevertheless reported that there seemed to be no insurmountable obstacle preventing the carrying through of three regulations: that there should be a net *selling* price for all books: viz. for net books the full published price; for non-net books at 6s. and under, a discount of threepence in the shilling; for books above 6s. a discount of twopence in the shilling; that all invoices should bear a notice that the books were supplied on condition that they were not sold to the public below the net selling price and that, in the event of an infringement, the full selling price of that book would be charged to the bookseller at settlement; and that old or dead stock would be exempt.

During the autumn the three regulations proposed by the conference were circulated to members of the Association, criticisms from six firms

[1] *The Athenaeum*, 11 December 1897.
[2] The Council also received at this meeting from W. Cunningham, the Cambridge economist, a copy of his letter to the Society of Authors resigning his membership.

only being received, and proposals for their application were developed by the President, John Murray. On 16 January 1899 the resulting scheme was adopted at a special General Meeting, subject to its approval by the Associated Booksellers and the regional societies, and since it formed the foundation for the first Net Book Agreement it deserves to be recorded *verbatim*. The provisions were as follows:[1]

(i) That a more general trial of the net system is advisable, especially in the case of books above 6s.: and that (*a*) New books should be issued as far as possible at net prices: (*b*) Existing books should, where it is practicable, be converted into net books, by taking off one-sixth of the present price.

(In the case of books thus converted to net prices royalties to authors would of course remain as now: e.g. if 2*d*. in the shilling is paid on a 12*s*. book, the royalty would remain at 2*s*. though the published price is reduced to 10*s*.)

(ii) That in the case of net books, the same trade terms as are now offered should be given to those booksellers only who agree to sell them at full price, and to those wholesalers who agree to enforce similar conditions with their customers; and that to others, net books should only be supplied at the full published price.

(iii) That a form of agreement between publishers and booksellers who assent to these proposals be drawn up, along with a list of the publishers thus assenting, and that this form be circulated by the Publishers Association among the booksellers, for signature by them.

(iv) That booksellers refusing to accept this agreement shall be treated alike by all the assenting firms of publishers.

(v) That the foregoing recommendations shall not apply to *bona fide* remainders or to dead stock.

(vi) It is understood that the Publishers Association cannot undertake the responsibility of detecting instances of underselling. This scheme is drawn up in the interest of the booksellers, as urged by the majority of them and their representatives, and it is clearly understood that the *onus probandi* in case of default rests with the various Booksellers' Associations.

A conference with the Booksellers' Associations followed, and the scheme was forwarded to the Society of Authors, and when the Association's annual meeting was held in March the adoption was confirmed.

THE NET BOOK AGREEMENT

During the summer of 1899 the form of the agreement was passed by the Association's solicitor, and its working details were determined. No discount was to be allowed for large orders and while existing contracts to supply, for example, Public Libraries at a discount were to be honoured, no fresh ones were to be signed; and libraries such as the 'London' were

---

[1] *Report of the Council for 1898–9*, p. 4.

to be supplied only through booksellers. By November out of 1,270 agreements sent out to booksellers 1,100 had been signed, but some London firms were still in opposition and the wholesalers, notably Messrs Simpkin Marshall in London and Messrs Menzies in Scotland, had not yet signed; and when the Publishers Association met again in a special General Meeting in that month it agreed that all members should sign a statement to be sent to the non-signatories that from 1 January 1900 they would be supplied with net books only on the terms and conditions specified in the agreement. In December the wholesalers came in, and on 1 January the Net Book Agreement came into operation. In the Council's Report on the last year of the century the President was able to record that it had been necessary to apply the penal clause in five cases only. 'Many persons', he concluded, 'whose opinions on such a subject are entitled to respect, have declared that, notwithstanding the [South African] war, which has necessarily had a bad effect on the sale of books, the retail book trade is at the present day on a more solid basis and in a more satisfactory condition than has been the case for many years.'

THE INTERNATIONAL CONGRESS OF PUBLISHERS

Towards the end of the century the members of the Association, taking time off from its domestic problems, enjoyed themselves as hosts to a gathering of publishers from other lands. In 1896 the Cercle de la Librairie had convened an international gathering in Paris, but when the Association came into being in March the time available for organizing representation at this First International Congress of Publishers was insufficient. At the Second Congress in Brussels in 1897 William Heinemann, representing the Association, read a paper on 'La Localisation du droit d'Auteur'; and the invitation from the Council, which he tendered, to hold the next Congress in London was accepted. The Third Congress, with John Murray as President, was held on 7–9 June 1898 and the entertainments included a soirée in Guildhall, a banquet given by the Stationers' Company in their Hall, and a visit to the library and the State apartments at Windsor Castle.

2

1901-1908
Through the Book War

The first quinquennium of the new century was for the most part a period of consolidation, with progress recorded on existing problems; with some new ones arising, notably from the Education Act of 1902; and with hope of one needed improvement continuing to be deferred. The membership of the Association rose from sixty-five in 1900 to seventy-four in 1905; and the annual subscription stood at four guineas. Of the three 'founding' Presidents—C. J. Longman, John Murray and Frederick Macmillan—Murray's presidency was past, Macmillan[1] was in the first of his two years when the century began, and Longman was returned again in 1902 for two more years; and in 1904 there came a new name, Reginald J. Smith, K.C., of Smith, Elder & Co.[2] William Poulten continued as Secretary,[3] and was to continue until 1933.

COPYRIGHT

On 1 February 1901 the Council held a special meeting to express its sorrow at the death of Queen Victoria and its dutiful congratulations to King Edward VII; and when in March the President made his report to the Annual General Meeting he noted with satisfaction that the King's Speech, the first of the new reign, had included a promise that the law of copyright would be one of the subjects with which Parliament would be asked to deal. In December the Council passed a resolution requesting the Government to redeem this pledge, on the lines of the Bill passed by the House of Lords in 1900; and in March 1902 a deputation of representatives of the Association, the Society of Authors, the Copyright Association, and the Music Publishers' Association, with two Members of

---

[1] He [then Sir Frederick Macmillan] was President again in 1911–13.
[2] One of the leading houses of the Victorian era, publishing Thackeray, the Brontës, George Eliot, Trollope, and the original *Dictionary of National Biography*; amalgamated with John Murray in 1917.
[3] at a salary of £100 p.a., rising to £125 in 1905.

Parliament, was received by the President of the Board of Trade, Gerald Balfour, who assured them that if Canadian objections had been removed, the Bill would be introduced in the next session. In October a visit by Sir Wilfrid Laurier, the Prime Minister of Canada, gave the opportunity of a meeting with him and he was provided with a memorandum of the views expressed, for communication to the appropriate Minister in Canada. While the expectation of a new Bill continued, the Association presented to the President of the Board of Trade in May 1903 a memorial praying that provision should be inserted in it for the continuance at Stationers' Hall of the Register of Copyright Publications which had existed for the last 400 years, at first under the private regulations of the Stationers' Company and since 1710 in accordance with statutory provisions, and urging also that such a Register might make it unnecessary for His Majesty's Customs Office to maintain its own register, of the inadequacy of which the Council had been complaining in the previous February. But the memorial received an acknowledgment only from the Railways Department of the Board of Trade and the expected new legislation did not come, although in 1905 the Association was able to note with satisfaction that the new Australian Act, providing a complete codification for literary and artistic copyright, followed the lines of the English Bill of 1900.

The memorial called attention to 'the constant activity among printers and publishers of other countries in their endeavours to import pirated editions for sale in the British Dominions' and from 1902 the Association was taking action, by representation to the Foreign Office and the India Office, through the International Congress of Publishers, and through its own agents abroad, to stop the importation into the Indian States, Australia and Egypt of pirated editions and of American editions of books first published in the United Kingdom. Protection of copyright in Egypt required the unanimous consent of the six responsible Powers, to which the President in the Annual Report of March 1905 expressed the hope that the improved relations between England and France might contribute.

In the year 1903 the Association made a considerable grant towards the expenses of a member firm in legal proceedings brought against them by the editor and one of the contributors of an *Encyclopaedia of Sport*, which the defendants had commissioned. The decision of the House of Lords, reversing a judgment of the Court of Appeal, established a principle of considerable importance to the trade: that where the proprietor of an encyclopedia employs and pays another person to write an article as part of an encyclopedia which the proprietor is producing at his own risk and expense, the natural inference of fact from the employment and payment —no agreement in writing or express words being necessary to the assignment of copyright—is, in the absence of evidence to the contrary, that the proprietor of the encyclopedia acquires the copyright in the article.[1] With its interest in case law thus aroused, the Council proceeded to commission E. J. MacGillivray to prepare a periodical record of legal decisions of interest to publishers, and the first of the series, *Copyright Cases*, covering the years 1901–4 was issued to members early in 1905.[2]

THE INTERNATIONAL CONGRESS OFFICE

The International Congress of Publishers held its Fourth Conference in Leipzig in May 1901, when a decision was made to set up a permanent office. The office, towards the expenses of which the Association paid initially an annual subscription of £100, was set up in Berne; and in a report in 1903 it was able to state that it was in communication with the postal authorities of all the countries represented in the Congress about book post rates; and that it had made representations to the Ministries of Foreign Affairs against continuing Customs duties on books and had received answers—encouraging for the most part—from fifteen countries including the United Kingdom. It noted that Britain was one of only three countries which had officially abandoned the misuse of the word 'edition' and had clearly defined 'edition' and 'impression'; but on the other hand it regretted that our Association did not consider it expedient to petition its Government to introduce the metrical system. The question of 'discount' to the public was to be the subject of a separate report, and information was being collected from the participating countries concerning the special regulations that might be in force governing the relations between authors and publishers, and on the supply of books to booksellers 'on sale or return'.

THE EDUCATION ACT OF 1902

The Education Act of 1902 considerably affected the manner in which the trade in school books had been carried on. In the place of 2,568 school boards it established 328 local education authorities. County Councils or County Boroughs now dealt with education as they dealt with other public services, aiding voluntary schools as well as the former board schools from the rates and being empowered to establish 'grammar type' secondary

---

[1] Lawrence and Bullen *v.* Aflalo [1904] A.C. 17.
[2] MacGillivray continued to prepare the series until 1949, assisted in the last volume by J. G. le Quesne.

schools. Whereas hitherto, with the exception of the London School Board which had been supplied at trade terms since 1875, the Managers of the school boards had generally bought school books from local booksellers, the new authorities were disposed to place large contracts for the supply of books to all the schools in their areas. Two problems followed: some authorities showed a desire to deal with publishers direct; and the power of the L.E.A.s and the size of the orders which they were able to place created severe competition among booksellers and, in particular, the school supply houses. At its meeting in March 1903 the Council, having received from the Council of the Associated Booksellers a request that publishers who might be approached by the new authorities should refuse to supply direct, itself resolved that it was inexpedient for publishers to supply books to the education authorities except through the retail trade; but at the Annual General Meeting later in that month it was agreed that in the first instance a committee should be set up to deal with questions affecting educational publishing and, as its first business, to consider the situation created by the Act of 1902. The Educational Books Committee[1] proceeded to canvass the views of such large educational publishers as were not then members of the Association[2] and, finding that most of these firms wished to keep a free hand, had to content itself with a recommendation that as far as possible the existing channels of supply should be maintained and that any serious blow struck at the interests of local booksellers would be a misfortune to the trade at large. The Council could do no more than send in June a circular to its members and to some educational publishers who were not members, requesting them to notify it before opening an account with any new education authority; and in the Report on the year 1904–5 it continued to advocate the employment of booksellers by the education authorities rather than a resort to publishers direct. Although the Council sought thus to retain the business with the L.E.A.s for the booksellers, it thought it undesirable to deny the authorities the benefit of unlimited competition and in July 1903 it rejected with sympathy a request from the Associated Booksellers that the discount allowable on the non-net prices of school books should be limited to $33\frac{1}{3}\%$.[3] It still had, however, a subsidiary problem to face: that

[1] The Committee became in effect a standing Committee and was thus the embryo of the later Group III.

[2] Notably, Sir Isaac Pitman & Sons, who became members in 1903, J. M. Dent & Sons, who had resigned in 1901 and did not rejoin until 1926, and William Collins Sons & Co. and McDougall's Educational Co., who were elected in 1906 and 1908 respectively.

[3] Then, and later, it was not unusual for the school supply contractors in particular to supply these books at a loss, in order to secure the schools' orders for stationery.

several of the school supply houses were themselves publishers and were inclined to give larger discounts on their own publications than on those of other publishers. Although the major supply houses were willing to discontinue this practice, the continuing contracts by which they were bound delayed the ending of it, but by October 1905 they had bound themselves by signature of what came to be known as the Educational (or non-net) Books Agreement not to give preferential discounts.

THE NET BOOK AGREEMENT IN OPERATION

During its first quinquennium the maintenance of the Net Book Agreement was not seriously threatened. During 1901 and 1902 the Council took successful action against petty infringements, at least some of which were accidental, and to stop the allowance to Public Libraries of excessive discounts on non-net[1] books to balance the charging of full prices on net books; and in November 1902 it declined to receive a deputation from the Library Association on the subject of the allowance of a discount on net books. The N.B.A. was also held to cover sales by exporters, and complaints that discounts were being allowed to American libraries were referred to the Trade Terms Committee in February 1904. A request in May to the American Publishers Association for assistance in maintaining English net prices received, not surprisingly, the reply that it was powerless to control the prices of books not copyrighted in the U.S.A., but nevertheless in November the leading exporters came out strongly with the advice that net prices could be maintained. In May 1904 an application by His Majesty's Stationery Office for recognition as a bookseller was, at first sight, rejected, but after a statement by the Controller that a large proportion of the books bought were resold to Government Departments, recognition was given in the following month. It may also be noted that in November 1905 the N.B.A. was extended to cover the sale of maps.

CHALLENGED BY 'THE TIMES'

Until 1905 the Net Book Agreement was subject to only these minor infringements, and it appears also to have been quickly accepted by the public as necessary for the maintenance of good bookshops. But in the summer of that year *The Times* instituted a new bookselling experiment which developed into a major attack on the Agreement, and the bitter ensuing struggle, which came to be known as 'the Book War', was to last for almost three years.

[1] It should be borne in mind that publishers were continuing to issue at non-net prices not only school books, but also some works of general literature including, in particular, novels.

Towards the end of the last century *The Times*, with a daily circulation insufficient to provide dividends satisfactory to its proprietors, had launched a number of subsidiary publications and in particular had entered into agreements in 1898 with James Clarke & Co.—a Chicago firm which specialized in reprinting and selling by direct methods books for which the ordinary demand had ceased—for the publication of a reprint of the 9th edition of the *Encyclopaedia Britannica* and in 1899 for a new 10th edition. Clarke, at the instigation of one of his salesmen, H. E. Hooper, had acquired the rights from A. & C. Black and set up an English company, in which Hooper and W. M. Jackson, of the Grolier Society, had shares and of which in 1900 they became the sole owners.[1] By a further agreement in 1904 *The Times* retained the services of Hooper and Jackson also as advertising consultants, to promote the circulation of the paper and the obtaining of advertisements in it; and they proposed to Moberly Bell, the Manager of *The Times*, the establishment of an office in the West End at which advertisements would be taken, theatre tickets sold, and books sold. The paper was to be offered to annual subscribers at £3, instead of the regular price of £3. 18s.;[2] and subscribers were to be offered, without charge, membership of a book club and library from which they could borrow books of their choice and purchase at a discount any which they wished after reading to keep. As the historian of *The Times* has written: 'the essence of all Hooper and Jackson's schemes was to establish direct contact with the purchasers', and

the Americans, accustomed to the more elastic mechanism of the trade in their own democratic country, and habituated to their own consistent plan of making the most direct contact with the actual book-buyer, thus enabling him to buy more books, never appreciated the determination of the English publishing trade to preserve the existing bookselling organization and the discount it lived on. To Hooper and Jackson the English discount system was a mere device to raise the price of books.[3]

It was anticipated that the new project might double the circulation of the newspaper and for that purpose the promoters were prepared to spend £100,000. 'No circulating library' it was said,[4] 'has ever been established with the deliberate object of spending money instead of making money'; and it was thus with a project which was not motivated by any sense of

---

[1] I am indebted to J. D. Newth, of A. & C. Black, for correcting the account which appears in *The History of 'The Times'*, III (1947), 443–6.
[2] The discount was withdrawn in 1905.
[3] *The History of 'The Times'*, III, 447–9.
[4] *ibid.* pp. 831–2, where the course of the Book War is told in some detail.

responsibility towards their welfare that publishers and booksellers had to contend.

During the first half of 1905 *The Times* approached the leading publishers individually, proposing that it should have the right to buy all books in any number from one to 10,000 copies at the best trade terms and that the publisher should undertake to spend in advertising in *The Times* not less than one-fifth (afterwards reduced to 15%) of the invoiced value of the books so bought, and suggesting a five-year agreement; in return for which 100,000 copies of a general catalogue would be issued to the public. Some publishers assumed that there was no probability of their sales to the book club amounting to five times the amount which they had already spent on advertising in *The Times*, but others would have nothing to do with the advertising stipulation. Among the booksellers the possibility that *The Times* would get what seemed to them to be equivalent to an extra 20% discount aroused intense irritation, and not unnaturally a demand for the same terms was presented to the Council at a special meeting, the first of many, on 13 July 1905. During the course of that meeting it was reported that *The Times* had signed the N.B.A. and, with its fears thus allayed, the Council recommended that *The Times* should be supplied on the same terms as the largest retail booksellers. There was no uniformity in the terms agreed with publishers and most firms refused to pledge themselves for more than one year; but four or five made a five-year contract and this seemingly unimportant concession became later a source of much trouble.

The Times Book Club opened at 93 New Bond Street on 11 September 1905. Books were offered for sale to subscribers to the newspaper in different classes at varying discounts. In Class A absolutely new copies were to be obtained at 25% below the published price, excepting books published at a price stated to be net. Class B comprised 'clean uninjured copies', 'virtually as good as new', their condition indicating that they had been in circulation about a month; and the discount was 20% on net books, and 35% on others. Class C represented a condition indicating about three months' use and Class D a circulation of about six months; and the discount on net books in these two categories was $33\frac{1}{3}$% and 50% respectively. It will be seen that the meaning of 'second-hand' was not clearly defined and although the other circulating libraries bound themselves not to offer net books as second-hand at reduced prices within six months of publication, the Book Club would only accept a definition of a second-hand book as 'one that had been used by more than two subscribers and returned in such a condition that it could not be sold as new'. This definition the Council was disposed to accept on the supposition that it would be reasonably interpreted, but remonstrances to the Book Club about apparent sales of new copies of two recent net books, Winston Churchill's biography of his father and a 2s. edition of *Lorna Doone*, led Edward Bell[1] (of George Bell & Sons) upon becoming President of the Association in March 1906 to arrange a meeting with Moberly Bell. The meeting took place on 9 May, both sides being accompanied by shorthand writers. Moberly Bell and Hooper, who represented the Book Club, still refused to accept the six months' time limit, but an understanding was reached that if the Book Club found themselves left with unsaleable stock they would offer it to the publisher or ask the Booksellers' Association to relieve them of it before they offered it to the public. Rather naïvely, as it now seems in view of the very large orders placed by the Book Club, the representatives of the publishers saw themselves as acting only as intermediaries between the Club and the retail trade rather than as meeting a major attack on the principles of the trade from a new irresponsible entrant which, in a brochure issued in the previous month, had announced itself as already the largest buyer of books in the world and as having been established with the sole object of increasing the circulation of *The Times* and therefore able 'to offer greater facilities than other libraries because it can afford to lose money while they must make it'.[2] 'So far as we were concerned,' writes Edward Bell, 'we considered the matter disposed of; and I must lay particular stress on this, because *The Times* afterwards took the line that we had acted in bad faith, and had suddenly broken off negotiations which had promised to lead to an amicable adjustment.' That the meeting proved futile is accounted for partly by the fact that when the President of the Booksellers' Association, Henry W. Keay,[3] who had attended the meeting, reported to his Association, it promptly declined to give any assistance to the Book Club by relieving it of surplus stock; and partly by a provocative development on the part of *The Times* of which Edward Bell and his colleagues must have been aware, but which they do not seem to have challenged at the meeting on 9 May. What we should now call the 'phoney' phase of the war was over and a bitter, distrustful struggle began.

[1] A narrative of 'the Book War' of 1906-8 written by Edward Bell in January 1910 was printed by Frederick Macmillan as an appendix to his account of *The Net Book Agreement 1899* (1924), and together with *The History of 'The Times'*, III, it has been a principal source.
[2] *The History of 'The Times'*, III, 832.
[3] He was President of the Associated Booksellers throughout its first twenty-five years.

THE BOOK WAR

The popularity of the Book Club had necessitated a move to larger premises, and to inaugurate the opening of the new building in Oxford Street there appeared in *The Times* of 1 and 17 May advertisements,[1] occupying two-thirds of a page and a whole page respectively, of 'the greatest sale of books that had ever been held', which would be open to all, whether members of the Book Club or not. The advertisement of 1 May stated that a stock of 600,000 books of all kinds, of which the published prices would amount to £222,000, was to be sold for less than £25,000; and it continued:

> Thousands of copyright books published at prices ranging from 30s. to 2s. 6d. each are offered NEW for 3d., 5d., 7d., 9d., 11d., 14d., 18d., 23d. and 30d. each...
>
> It is the opinion of *The Times* that books have always been sold at too high a figure, that if their prices were reduced to a scale more in correspondence, for example, with the price of a newspaper, books would circulate in correspondingly larger numbers. In the course of two days, for instance, *The Times* itself prints in its columns and sells for 6d. (the price of two issues of *The Times*) as much news matter as is contained in an important biography published at 21s. net. The cost of the paper, printing and binding in the case of such a book may amount to 1s. 6d. The enormous balance of its price goes in profits to the publisher, author and booksellers, wholesale and retail. An enormous balance indeed, but by no means an enormous profit, because the quantity sold is small. At our great sale such books will be sold for 23 and 30 pence. Where do the books come from? *The Times* has purchased from the publishers direct 500,000 books of different kinds; it has secured a further stock of 70,000 from dealers and other libraries, and has added to them books from the surplus stock of *The Times* Book Club. The books acquired direct from the publisher are of course entirely new.

On the opening day (1 May) the building was so besieged by crowds that, as *The Times* reported on the following day, the doors had to be closed soon after 10 a.m. and reopened only at intervals of an hour for two or three minutes at a time. It will be seen that there was no pretence in the advertisement that all the books were second-hand and the long list of 'sample bargains' included in the advertisement of 17 May was headed by the biography of Lord Randolph Churchill, published at 36s. net and offered at 84d. [*sic*], and contained a number of recent 6s. novels at 9d. or 11d. For one element in this great jumble sale the publishers were

---

[1] Edward Bell (*op. cit.* p. 42) does not mention the advertisement of 1 May and surely is wrong in implying that he and his colleagues knew nothing of the sale until 17 May, eight days after the meeting.

responsible, for the Book Club had been making large purchases of remainders from them.

By the Net Book Agreement, which Moberly Bell had signed, the sale of second-hand books was not limited either by definition or in time, but as the historian of *The Times* concedes[1] 'the term "*bona fide* remainders", which it did contain, could hardly have been intended to apply to books recently published'. If *The Times* was not prepared to observe the spirit of the Agreement, the whole of the trade was now determined to preserve it in the letter, and on 4 July a special General Meeting of the Association was held to consider resolutions proposed by John Lane. Lane first moved that:

(*a*) a net book shall not be sold (as a second-hand book) by libraries or booksellers under its published price within six months of its publication;

(*b*) a book subject to discount shall not be sold to the public, new or second-hand, at less than 75% of its published price within six months of its publication.

School books are excepted from the above rules.

As an amendment to (*a*) it was moved:

I. that second-hand copies of net books shall not be sold at less than the published price within six months of publication;

II. that new copies of net books shall not be treated as dead stock within twelve months of the date of purchase nor shall at any time afterwards be sold at a reduction without having been first offered to the publisher at cost price or at the proposed reduced price whichever is the lower.

Both parts of the amendment were passed, the second after two further amendments had been negatived; but section (*b*) of Lane's resolution was referred to the Council for consideration and report within five months.

Lane also moved a resolution to prevent misunderstanding between publishers and booksellers about remainders: 'when a book is remaindered within two years of its publication, the publisher shall make the bookseller an allowance of the difference between the amount originally charged and the remainder price. If, however, a publisher remainder a book two years or more after publication, he shall not be responsible for copies which remain unsold upon the booksellers' hands.' But a motion for 'the previous question' was passed and the matter was shelved. The meeting ended with a speech by John Murray in which he put forward the suggestion, amidst general applause, that publishers should abstain from advertising in *The Times*.

[1] *ibid.* p. 832.

The Council immediately took in hand the preparation of a revised Net Book Agreement, which was to supersede the existing one on 1 October except that those few houses, notably Longmans and the wholesalers, Simpkin Marshall & Co., who had contracts for a term of years with *The Times*, were to honour them and during their continuance to supply under the terms of the old Agreement, if *The Times* refused to sign the new one. On 30 July the new Agreement was communicated to Moberly Bell. He declined to accept it, without a year's previous notice, on the ground that *The Times* was under contract with their subscribers to supply them on the terms advertised. The Council offered to consider the new terms as applicable only to new or renewed subscriptions, but Moberly Bell again declined. The correspondence between the two sides since the meeting on 9 May was subsequently printed in *The Times* of 3 November in a letter in which Edward Bell and Longman refuted charges of bad faith. *The Times* believed that the Association had allowed the Book Club to go on, lulled into false security, and had then on 30 July dealt it a smashing blow; the Association conceded that the formula for defining 'second-hand' had seemed to it acceptable, but claimed that the Book Sale had shown that *The Times* did not mean what it said.

The Times took the offensive in four conspicuous advertisements in the issues of 25 September and the three following days. In the first it charged the Association with attempting to establish a restrictive monopoly which ran counter to the natural course of business and public convenience and affirmed its determination, by large buying and the consequential need to sell off in particular Class B books after about one month's use, not to keep subscribers waiting as other libraries did. The second was devoted to 'the real evil' of publishers' excessive prices and the smallness of the discount allowed by them to booksellers: lower prices and better treatment of the booksellers would enable publishers to print many more copies. Ignoring the author's royalty, the publisher's overhead expenses and advertising expenditure, and the discount to booksellers, the advertisement claimed that a book published at 36s. cost only 4s. to produce and thus gave the publisher a profit of 800% on his outlay: a charge which, as Edward Bell noted in his subsequent narrative, might seem incongruous to the readers of a paper which cost three times as much as any other London daily. In the third advertisement the 'paucity of bookshops compared with the plethora of shops in which other commodities may be bought' was attributed to the publishers' high prices and low trade discount and was held to support the Book Club's policy of price-cutting. In the final advertisement *The Times* championed all booksellers who dealt in new books against resolution II passed at the meeting on 4 July; and, aware perhaps that John Lane's resolution about 'remainders' had been shelved, attacked regulations under which 'the publisher may issue a cheap edition of a new book whenever he likes, selling off as a "remainder", at any reduction which suits him, such stock as he may himself still hold of the expensive edition, paying no attention whatever to similar copies which remain unsold on the booksellers' shelves'.

The advertisements were answered by the Association in a letter which was printed in *The Times* of 10 October; and at a special General Meeting on the same day it was agreed not to supply the Book Club with net books, except at the full published prices, not to sell it 'remainders', and to abstain from advertising in the paper. It was agreed also to set up a Committee with a fighting fund at its disposal, to conduct the Association's case in the correspondence columns of *The Times* and other papers; and the Committee proceeded to employ James Douglas of the *Evening News* to draft for its approval the many letters which were to appear over the signature of the Secretary in the next six months. As Edward Bell writes:[1]

the dogs of war being thus let loose, a terrible amount of barking and snarling ensued. It would have been better, and have saved much trouble and expense, to have left *The Times* a monopoly in invective, and to have limited our contributions to the dispute to a reasoned statement of our position, and to a quiet correction of any absolute perversions of the truth.[2]

The Times was fighting on its home ground and although it scrupulously printed every letter from the Association—and, except on one occasion, without any omissions—it was easy for it to have the last word editorially and difficult for the Association to get across the technicalities of the publishing business. *The Times* also commissioned an economist, Arthur Shadwell, to write a series of seven articles, which appeared between 16 October and 20 November, and when the Association found his articles lacking in understanding of the true merits of the case, it was partly its own fault, for doubting his impartiality it had refused to give him a statement of its case.[3]

Throughout the correspondence *The Times* printed at the head of the column the following epitome of its case:

The only question at issue between *The Times* and the publishers is as to whether the Book Club shall or shall not be allowed to sell second-hand 'net'

---

[1] *ibid.* p. 46.
[2] Such a statement, drawn up by John Murray, was issued as a pamphlet entitled '*The Times*' *and the Publishers.*
[3] He did nevertheless demonstrate the fallacies in the argument that publishers made an 800% profit on their outlay.

books before they are six months old. *The Times* Book Club maintains its right to sell *bona fide* second-hand books when it likes and at what price it likes. The Publishers try to prohibit this, and to enforce their prohibition by charging higher prices to *The Times* than to other purchasers and by withdrawing their advertisements from *The Times*.

To that the Association replied that it was obvious that the whole controversy turned upon the meaning of '*bona fide* second-hand'. In order to arrive at a reasonable definition the whole trade had been consulted; a large proportion of those engaged in it desired that net books should not be sold second-hand within twelve months but there was absolute unanimity among publishers and booksellers that six months was the minimum. The Book Club alone refused to accept this definition and claimed the right to sell second-hand books when and at what price it liked. It also, the Association maintained, was selling books as second-hand which were not made second-hand by use; for example, Moberly Bell had stated three days after the publication of Bram Stoker's *Irving* that it would be sold at a reduction of 20% at the end of the week. If the Book Club wanted to give a bonus, it must pay for it itself by foregoing its trade allowance.

Day by day throughout the autumn of 1906 and into the winter *The Times* printed letters from supporters of both sides. While some authors attacked the publishers as a selfish trade combination opposed to cheap literature, in a letter of 23 October the Secretary of The Society of Authors reported that the managing committee of the Society had resolved that the course being pursued by the Book Club was opposed to the interests of authors and gave its considered reasons for supporting the two trade Associations. On the other side the issue of 3 December contained a memorial, organized by Henniker Heaton, M.P. (known for his advocacy of Post Office reform) and signed by some 10,000 persons headed by the Lord Mayor of London, which protested against the publishers for selling books to people outside the United Kingdom at about half the price charged to people within it,[1] against the rule which prevented the public from buying good books at low prices second-hand, and generally against any interference with the rights of a man to dispose of his property on what terms he pleased.

The progress of the dispute was followed also by other newspapers and periodicals, the majority of which seem to have been against *The Times*. *Truth* on 3 October, in an article entitled '*The Times* on the Warpath',

---

[1] Referring to the practice of issuing cheap editions of 6s. novels for colonial circulation, at 3s. 6d. or 2s. 6d. in paper covers.

gave a detailed account and followed it a fortnight later with a satire on 'Book Clubs and Breeches Clubs';[1] and a balanced exposition appeared in *The Spectator* of 13 October. '*The Times*,' it wrote,

no doubt, alleges that it had always kept strictly within the letter of the law as to the agreement which it signed in regard to the sale of 'net books'. A little reflection, however, will show that it kept the letter rather than the spirit of the agreement, or, to put it another way, that the original agreement was so loosely worded that its spirit could be defeated by the sale of so-called second-hand books at very large reductions. *The Times*, no doubt, had a perfect legal, and therefore a perfect moral, right to manage its own affairs in its own way, and to sell books bought under the net agreement in any way that did not legally infringe that agreement. The booksellers, on the other hand, had an equally good right, when the old agreement expired, to make a new form of agreement as to net books.

And it concluded: 'we cannot resist expressing what we are sure is in the minds of thousands of Englishmen today—namely, that *The Times* had much better stick to its last, and be the best newspaper in the world, rather than turn itself into a cheap lending library and a bookseller at cut-throat prices.'

During November and December 1906 an attempt at mediation was made by a number of eminent authors. Following a letter to *The Times* on 26 October from Mrs Humphry Ward, whose husband was one of the paper's leader-writers, Lord Goschen formed an unofficial Committee including Laurence Binyon, Hall Caine, Mary Cholmondeley, Sir Norman Lockyer, E. V. Lucas, G. W. Prothero, Elizabeth Robins, G. M. Trevelyan and Mrs Humphry Ward, with Henry Newbolt as honorary secretary; and at a meeting on 6 November the Committee formulated, and communicated to the Association and the Book Club, resolutions that no new copies of any books should be sold at a discount within six months of publication and that second-hand copies should not be sold within three months and then only to subscribers to *The Times*. Although these suggestions reduced the intervals laid down in the new Net Book Agreement and made no distinction between net books and others, the Council of the Association sought and received from Lord Goschen's Committee permission to consult the Society of Authors officially, and on

---

[1] A sample may be quoted: 'Having the highest example in journalism before me, suppose that I decide to start a TRUTH TROUSERS CLUB. I offer to all annual subscribers to TRUTH on the usual terms the following privileges, in addition to the delivery of the paper weekly: (1) the use of a pair of two-guinea trousers free; (2) the right to change their trousers as often as they like at this office; (3) the option of purchasing club trousers at reduced prices in Class A, B, C or D, according to the state of repair that the garments are in after the other members have had a turn at them. In Class A the trousers will be undistinguishable from new; in Class D these will be a bit baggy at the knees...'

22 November a meeting was held with Sir Henry Bergne and Douglas Freshfield and with Henry Keay representing the Associated Booksellers. A compromise was agreed: that net books might be sold second-hand five months, and novels four months, after publication; and that unsaleable new copies of net books might be sold six months after purchase, but must first be offered to the publisher at trade price or the proposed reduced price, whichever were the lower. But although Keay with some difficulty persuaded his colleagues to accept this as a modification of the N.B.A., Moberly Bell rejected it and Lord Goschen's effort came to an end.[1] No greater success attended an indirect attempt at conciliation by the Prime Warden of the Fishmongers' Company, at whose livery dinner on 13 December Moberly Bell and Edward Bell, responding to the toast of 'The Diffusion of Literature', expressed their divergent views.

In its increasing efforts to close the channels of supply to *The Times* the Council had to keep in mind the possibility of indictments for conspiracy in restraint of trade and it engaged the services of Messrs Lewis and Lewis as solicitors and Rufus Isaacs (later Lord Reading) as counsel. By the end of 1906 many of the leading publishers (except Longman's and a few others whose contracts with the Book Club were still current) had warned Messrs Simpkin, the wholesalers, that, notwithstanding their contract, they were not to supply the Book Club; and threatened on the one hand with black-listing by the Association and the discontinuance of trade terms, and on the other with an action for breach of contract by the Book Club, Simpkin chose the latter as the lesser of two evils.[2] Although, as the historian of *The Times* records, newsagents and amateur booksellers found profit in buying for resale to the Book Club and although purchases were made from booksellers on the Continent, the results 'were far from satisfactory. In fact the Book Club's promises and the hopes of early days were not and could not be fulfilled.'[3] The Book Club was indeed reduced to sending circulars to its subscribers asking them to refrain as far as possible from ordering the books published by the firms who refused supplies; and although the publications of these firms were reviewed with conspicuous impartiality in *The Times Literary Supplement*, the editor appended to the reviews the following note:

The publishers of this book decline to supply *The Times* Book Club with copies on ordinary trade terms, and subscribers who would co-operate with *The*

---

[1] The resolutions passed by the Committee on 6 November and the ensuing correspondence with the Association and the Book Club was printed in *The Times* of 14 and 17 December.
[2] An action against them was entered by *The Times*, but not brought to court.
[3] *The History of 'The Times'*, III, 833.

Times, to defeat the Publishers' Trust may effectively do so by refraining from ordering the book so far as possible until it is included in *The Times* monthly catalogue.

During the spring of 1907, the letters printed in *The Times* increased in quantity and, perhaps because the Association ceased to reply, the understanding of its attitude and the quality of the invective diminished; and *The Times* also ran a series of articles, based on an ingenious travesty of an article by John Murray in the *Contemporary Review* of December 1906, in which it returned to its exposure of the concealed profits made by publishers at the expense of authors and the public. Aroused by these articles the Society of Authors became divided, and at its Annual General Meeting in March considerable opposition to the support given to the Association by the Society's committee of management was led by Sidney Lee and Bernard Shaw; and when Shaw's publisher, Constable, continued to refuse to supply the Book Club with copies of *John Bull's Other Island* he took it out of their hands and produced copies bearing the imprint 'Issued by the author for *The Times* Book Club'.

There was a lull in hostilities during the summer of 1907, but the opening of the autumn publishing season gave fresh stimulus to the contest and was to have a decisive effect upon it. The great event of the season was the publication of *The Letters of Queen Victoria*, edited in three volumes by Lord Esher and A. C. Benson and published at three guineas by John Murray. What followed is laconically recorded by the historian of *The Times* in these words:

Though supplies could not be bought from the publisher in the usual way, a small number of copies was obtained and circulated by the [Book Club] Library. In a review of the book, *The Times* commented on the high price at which it had been published. More serious references to the price were made later in a letter to the Editor signed 'Artifex'. Mr. Murray bought an action for libel against *The Times* in which he afterwards obtained a verdict, and was awarded damages of £7,500.[1]

'Artifex', in fact, in two letters made a bitter attack on John Murray, using conjectural figures of the cost of the work, concluding that 'these figures spell simple extortion', and comparing Murray inferentially with Judas Iscariot. In the action, which was tried before Mr Justice Darling in May 1908, it was established not only that the figures were seriously incorrect and had damaged the sale, but also that—and here was the relevance of the case to 'the Book War'—the references to the high price

[1] *ibid.* p. 834.

in the review had been interpolated by Moberly Bell and that 'Artifex' was a member of the staff of *The Times* writing under Hooper's instruction.

By the end of 1907, the management of *The Times* had serious domestic difficulties to contend with. Not only were some of the smaller proprietors in the partnership critical of 'alien' book schemes, but the financial ability of the managers to continue the paper was in doubt. In January 1908 C. Arthur Pearson, who was negotiating for the purchase, invited leading members of the Council of the Association to meet him and a tentative arrangement for ending the Book War was reached. But by March *The Times* had been bought, at first anonymously, by Lord Northcliffe, and on 7 May he visited William Heinemann and Frederick Macmillan[1] to inform them confidentially and to seek their assistance in ending the war. Northcliffe said that while he wished to reach a settlement, if only to reduce the loss on the Book Club, which he put at £800 a week, Moberly Bell and his staff still believed that publishers were a wicked race of men, victimizing the reading public, and might resign their positions if he gave in too openly; Heinemann and Macmillan for their part said that any negotiations which did not aim at preserving the letter and the spirit of the Net Book Agreement would be futile. Other meetings followed with Northcliffe and with his secretary under the assumed name of 'Mr Bates', and draft clauses of a settlement were approved by the Council at its meeting on 14 May. Nothing further happened until towards the end of August when Northcliffe wrote to Macmillan from Paris to say that his partner, Kennedy Jones, had full authority to make a settlement. After an abortive meeting on 2 September at which Kennedy Jones sought, and Heinemann and Macmillan refused, some concession from the draft settlement it was agreed at a final meeting on 14 September: that the Book Club would accept the Net Book Agreement without modification; that the Book Club would be at liberty to take orders for second-hand books to be supplied at reduced prices at the expiry of a 'close time' (the 'close time' for net books being six months from publication and three months for other books), but no such orders to be filled until the 'close time' had expired; that before selling surplus new copies of both net and other books at a reduction the Book Club would offer them back to the publisher at cost price and would be at liberty to retail them at reduced prices only in the event of the publisher's refusal to buy them. The Council of the Association, for its part, was to pass a resolution recommending all the members of the Association to supply their publications to the Book Club

---

[1] The subsequent events are recounted by Macmillan in *The Net Book Agreement 1899*, pp. 67-9.

on the best terms that they gave to any other bookseller or librarian; and both parties agreed that they would make no public announcement of the settlement. Although some members of the Council demurred strongly at its meeting on 17 September against the recommendation for the supply of the Book Club on the best terms, Edward Bell was authorized to sign the agreement; and Moberly Bell signed it for *The Times* on the following day.

'The Book War' was over; publishers' advertisements began to appear again in *The Times* and the *Literary Supplement*, and reviews were no longer accompanied by the note about the 'Publishers' Trust'. For *The Times* it had been a struggle not primarily to benefit the public by lowering the prices of books, but to increase the circulation of the paper; for the book trade it had seemed a matter of life or death. 'Failure' wrote Frederick Macmillan,

would have endangered and probably broken down the Net Book System which I inaugurated seventeen years ago, which was adopted by the Publishers Association and the Associated Booksellers ten years ago, and which has saved the Retail Book Trade throughout Great Britain from bankruptcy and the virtual extinction with which it was threatened, owing to the suicidal system of under-selling.[1]

The Net Book Agreement was not again to be seriously challenged for more than fifty years.

It remains to record an amicable sequel. At the beginning of October 1908 there appeared in *The Times* (and other morning newspapers) an announcement, occupying a full page in large type, that:

The King, being aware of the great interest taken by the Nation in general in 'The Letters of Queen Victoria' recently published, has commanded that a new and popular edition should be brought within reach of all His Majesty's subjects. This edition will be in three volumes, crown 8vo, and will contain 16 illustrations, as well as the complete text of the larger work carefully revised; it will be sold at 6s. net, bound in red cloth. It will be published by Mr. Murray in conjunction with *The Times*. The book will be obtainable at all booksellers, including *The Times* Book Club.

CRITICISM OF THE ASSOCIATION

The conduct of the dispute by the 'Association' had not been without critics among its members. In a confidential memorandum, of which there survives a copy[2] sent apparently to A. D. Power, of Sir Isaac Pitman &

---

[1] Macmillan, *op. cit.* p. 77.
[2] London School of Economics Library of Political and Economic Science, Coll. G.330.

Sons,[1] and dated by him 12 September 1907, John Lane put forward 'Tentative Suggestions for extending the scope of The Publishers Association'. Arguing from the greater effectiveness of the Authors' Society and the Associated Booksellers, and from the new power of the literary agents, he claimed that 'The Publishers Association seems not only to lack initiative but also the most rudimentary organization. If proof be wanted of this, we have only to turn to *The Times* Book War. The action taken by The Publishers Association was not spontaneous; it found itself urged to fight by the action of the booksellers'. To meet 'some of the obvious criticisms' of the limited activities of the Association he put forward the following proposals: that the Association should acquire in central London an office and a showroom where a copy of each book published by members would be on view, say for one year, to the trade and the public and where orders would be taken for supply through booksellers who were members of the Associated Booksellers; that the Association should issue a weekly catalogue for distribution at home and abroad; that a copy of any book at 5s. or more might be obtained on approval by any bookseller or librarian, provided that the order came from a bookseller who was a member of the Associated Booksellers; and that arrangements should be made with the chief daily and weekly papers to insert a note in their publishers' advertising columns to the effect that all the books announced might be seen at and ordered from the Association's office. Thus, the smallest publisher would no longer be at a disadvantage with the largest, and every publisher and bookseller would have to join his respective association. 'All this' Lane concluded, 'would tend to bring about the esprit de corps which is so much needed.' Lane's scheme, which was surely ahead of his contemporaries' ideas about the functions of the Association, did not receive official consideration. Although, as we shall see in the next chapter, under the stress of war-time conditions the Council took a temporary hand in sales promotion, it was to be many years before direct assistance to members in the selling of their publications was accepted as a legitimate function of the Association.

[1] Subsequently of W. H. Smith & Sons.

3
1908-1914
Copyright and novel prices

The promise of legislation on copyright which the King's Speech had contained in 1901 was not fulfilled within King Edward's reign. For almost seventy years the basis of the law for books and music was the Act of 1842, supplemented by an Act of 1862 which gave protection for the first time to paintings, drawings and photographs, and by more than twenty other Statutes which had been passed from time to time to deal with abuses as they arose. There was a great need for consolidation and simplification and for a reconsideration of the term of protection, which lasted only for forty-two years from publication or for seven years after the death of the author, whichever were the longer.

NEW COPYRIGHT LEGISLATION

International copyright was governed by membership of the 'Union Internationale pour la protection des œuvres littéraires et artistiques' and regulated by its Berne Convention of 1886 and the additional Acts of Paris of 1896 under which each member country undertook to give to the authors of other members the same rights as to its own, but to which not even all the countries of Western Europe had subscribed. In the spring of 1908 the Union gave notice of a Conference to be held in Berlin in October and submitted an 'avant-projet' for the revision of the terms of the Convention. After a study of these proposals, and of a memorandum by MacGillivray, by the General Purposes Committee[1] the Council warned the Board of Trade in April that some of the proposed changes in the Convention would not be in conformity with the existing state of the

---

[1] The General Purposes Committee, consisting of the past and present Officers, was first constituted as a formal Committee at this time, i.e. March 1908. In 1926 William Longman was added to it and subsequently the following also were elected: G. H. Bickers (Bell's), W. Farquharson (Murray), G. Wilson (A. & C. Black). From 1930 three additional members were co-opted for the consideration of matters relating to publication agreements only: G. O. Anderson (Harrap's), C. S. Evans (Heinemann's), Harold Raymond (Chatto & Windus). The Committee ceased to exist in 1937.

British law. New legislation, the Council stated, would be required to deal with the following proposals:

(1) The extension of copyright to 50 years after the death of the author, which would assimilate our term to that of France, Belgium, and several other countries.
(2) The placing of translations on the same footing as the original works.
(3) The protection of musical airs from mechanical reproduction.
(4) The securing of copyright without unnecessary formalities and the abolition of imposts of any sort on books.
(5) The promotion of strict reciprocity between this country and all others which are not signatories of the Berne Convention.[1]

When the Conference met in Berlin the changes proposed, including the protection of cinematographic productions, were embodied in a new Convention to be ratified by the legislatures of the member countries by 1 July 1910, but the British delegates could not of course promise concurrence.

The consequential changes necessary in the British law were referred to a Departmental Committee of the Board of Trade, of which Sir Frederick Macmillan was a member and before which William Heinemann, who became President of the Association in April 1909, and John Murray and Reginald Smith gave evidence. But, as members were warned in the Annual Report in March 1910, progress towards new legislation was seriously postponed by the constitutional crisis which had resulted from the rejection of the Budget by the House of Lords in November 1909 and by the possibility, after the general election of January 1910, of their continuing recalcitrance, which was not removed until the following April. Then in May King Edward died. Nevertheless on 26 July a Bill was published, purporting to give effect to the recommendations of the Berlin Conference and to take account of the wishes of the self-governing Colonies. A Copyright Committee[2] of the Council was set up in August, to consider the provisions, with the assistance of MacGillivray, and to confer with the Music Publishers' Association, the Copyright Association, and the Society of Authors; and a joint deputation put their views before the Board of Trade, particularly with reference to Imperial copyright. Some of the advantages resulting from Berlin, it was feared, were to be

---

[1] *Members' Circular*, no. 1 (May 1908), p. 2: the first number of the *M.C.*, issued at first irregularly, from 1916 monthly except in August and September, and from 1922 monthly except in August. A complete set of the *Members' Circular* is deposited, under agreed conditions, in the British Library of Political and Economic Science at the London School of Economics.

[2] The President, Edward Bell (Vice-President), Sir Frederick Macmillan, Murray, Reginald Smith.

offset by considerable sacrifices: while it was the wish of the Government to induce the Colonies to adopt legislation practically uniform with the new Bill, nevertheless they were to be given absolute liberty to legislate as they chose, and the Council feared that colonial copyright might become 'a doubtful and precarious element', with definitions of publication and modes of registration varying from one part of the Empire to another. Another feature of the Bill which disquieted the Council was the division of the term of copyright into two periods, the first of unrestricted ownership being followed by a second period in which permission for reproduction could be given by the Comptroller General of Patents; and the Council urged that no application to the Comptroller General should be allowed until after the expiry of a period equal to the then existing term of copyright. On the other hand the Council was gratified to note that information which it had given to the Board of Trade about the use of the cinematograph in a lecture in Germany and about the new Toussaint–Langenscheidt method of teaching languages by means of the gramophone had been followed by provision in the Bill for the protection of all copyright works, and not only musical compositions, from unauthorized reproduction by mechanical means.

On 30 March 1911 the Bill as amended was introduced into the House of Commons by the President of the Board of Trade, Sydney Buxton, and was referred to the House in committee; and the Council nominated Sir Frederick Macmillan[1] and C. J. Longman to serve on a Joint Copyright Committee with representatives of the Authors' Society and the Copyright Association, to watch the progress of the Bill through the House. At the first meeting of the joint Committee on 26 April Macmillan, who took the chair, was able to report that on the previous day he had seen Buxton, who had told him that the duration of the period of copyright which he now proposed to suggest would be the lifetime of the author and fifty years thereafter, with the proviso that during the last twenty years any person should be at liberty to reprint subject to notice of intent and to the payment of a 10% royalty. To this proposal the Committee unanimously agreed and made suggestions for its administration. The Committee was in frequent session throughout the summer, considering amendments as they arose in the Commons and making recommendations to the Government and, after the printing of the Bill on 13 July, making arrangements for the submission of further amendments in the House of Lords. In particular, at a special General Meeting of the members of the Association on 30 June, a resolution was passed protesting against the proposed

[1] Macmillan became President (for the third time) in March 1911.

addition of the National Library of Wales to the five libraries already entitled to receive free copies of books under the Copyright Act. The following extract from a memorandum which was circulated to members of the House of Lords may be of historical interest:

This exaction is the remnant of an enactment connected with the literary censorship, established after the Restoration and intended to prevent the publication of heretical, blasphemous, immoral or seditious books. After a number of years the statute expired and the delivery of the three copies then demanded ceased. At the beginning of the 18th century it was, however, revived, and in 1774 further copies were added for the English and Scottish Universities as well as for the public schools of Winchester, Eton, and Westminster. Some few years later Lord Colchester's Act for extending the laws of copyright to Ireland required two additional copies for the Dublin libraries, bringing the number up to eleven in all. By the Copyright Act of 1842 this number was reduced to five, one for the British Museum, one each for the Universities of Oxford and Cambridge, one for the Advocates' Library at Edinburgh, and one for Trinity College, Dublin. But this reduction was made only after it had been proved that the impost was a crushing one in the case of the publication of expensive scientific works; and an article which appeared in the *Quarterly Review* in 1819 cites a list of books, the publication of which the enactment either prevented or severely handicapped.

The memorandum went on to state that an examination by the accountants of certain publishing houses had shown that the cost of providing the five copies already required was equivalent to an additional income tax of threepence in the pound on the profits of their businesses, which with income tax at 1s. 2d. in the pound, as it then was, was a serious impost. The memorandum was also accompanied by a table showing the number of free copies necessary to registration under the copyright laws of 24 foreign countries: 4 countries, 3 free copies; 9 countries, 2 free copies; 11 countries, no free copies. The protest to the Lords, however, proved ineffectual and the new Library of Wales was added as a sixth library of deposit, although subject, unlike the other five, to regulations to be made from time to time by the Board of Trade.[1]

THE 1911 ACT

The Copyright Act 1911 received the Royal assent on 16 December and came into force on 1 July 1912. Although of necessity a compromise giving neither authors nor publishers nor the public all that they would have liked, it was a very great improvement and simplification of the

[1] In 1914 Sir Frederick Macmillan was appointed as the first representative of the Publishers Association on the Court of Governors of the National Library of Wales.

previous law. Its principal new features may be mentioned. It covered unpublished as well as published literary, artistic and musical work and new classes of work such as cinematograph productions and mechanical recordings; and translation rights and the rights of reproduction by mechanical instruments were reserved to the author. Registration of copyright was abolished. The duration of copyright in a work by a single author was the author's lifetime and fifty years thereafter with the proviso that after the lapse of twenty-five years from the author's death anyone could reproduce the work on certain conditions, in particular the giving of notice and the payment of a 10% royalty. Any 'fair dealing' with a work for the purposes of private study, research, criticism and review and the publication of 'short passages' in collections, 'mainly' of non-copyright material, for use in schools were permitted. Territorially the Act covered British possessions, except the self-governing territories, which were given a free hand in copyright legislation. Finally, it provided that no person should be entitled to copyright otherwise than under its provisions, so that the common law copyright previously recognized in unpublished works was superseded.

The interpretation of the clauses permitting 'fair dealing' in a work of scholarly research and the use of 'short passages' for use in schools was to cause some difficulty through the years. For the former the Council could suggest no substitute for the obtaining of permission from the copyright-owner. For the latter the Council also recommended that notice of intention should be given to the copyright-owner, but expressed the hope, in its Annual Report for 1913–14, that a general understanding would establish the definition of a 'short passage' as up to 1,000 words of prose and 100 lines of poetry, provided that in either case the extract did not comprise more than one-third of the complete poem, essay or story from which it was taken. In this definition the Society of Authors concurred. An elegant opinion by MacGillivray also advised members that the permitted use of such passages in collections 'mainly' of non-copyright material meant that the copyright extracts should be a comparatively insignificant amount, and not 49%, of the whole.[1]

The Act of 1911 came into force, without further legislation, throughout the British Empire including British India (but not the Princely States), but excluding the self-governing territories of Australia, Canada, Newfoundland, New Zealand, and the Union of South Africa. The Act itself or substantially identical ones were subsequently adopted by the four territories other than Canada, which therefore remained under the old,

[1] *Members' Circular* 1, no. 16 (February 1914), 136.

otherwise repealed, Act of 1842, so that British copyrights in Canada and Canadian copyrights in Europe continued to enjoy the shorter term and the lesser benefits of that Act.[1]

None of the countries which accepted the Berlin Convention of 1908 required manufacture within their territory as a condition of copyright, and all accepted 50 years from death as the period of copyright except Germany which adhered to her own period of 30 years. The signatories to the Convention were joined by Holland and Hungary after the holding of the 7th and 8th International Congresses of Publishers in their countries in 1910 and 1913 respectively; but hopes which the Council began to have in 1911 of an agreement with Russia are still unfulfilled. The United States continued to require manufacture in their country, but an Act of 1909 gave 60 days' grace to a book in the English language printed abroad. If a copy of the book was deposited in the Copyright Office within 30 days after foreign publication, *ad interim* copyright was granted for 30 days and if during that interval an edition was printed from type set in the U.S., then copyright was secured for the full term, which was extended from 42 to 56 years from first publication, i.e. for an initial period of 28 years, renewable by a further registration for another 28 years. For this alleviation members of the Association were largely indebted to negotiations conducted by William Heinemann during a visit in 1905.

THE INTERNATIONAL CONGRESS

Heinemann was also active in the Executive Committee and the Permanent Commission of the International Congress of Publishers, upon which he represented the Association from 1896 to 1913,[2] and at the Sixth, Seventh and Eighth Congresses which were held, respectively, in 1908, 1910 and 1913, at Madrid, Amsterdam and Budapest. Among matters dealt with were the formulation of needed improvements in copyright legislation,[3] which formed the basis of the Berlin Convention, and efforts to secure its acceptance by non-Union countries such as the South American republics; the observance of a code of practice in the dating of title-pages; and the

---

[1] An opinion by MacGillivray in 1915 confirmed that first publication in Canada or simultaneous publication in Canada and in the United States (e.g. of a work of American origin) secured copyright in no part of the British Empire except Canada.

[2] He was succeeded by G. S. Williams of Williams and Norgate.

[3] The Association's delegates, representing the principal sufferers from the American insistence on local manufacture, advocated strict reciprocity between signatories and non-signatories of the Berne Convention, but gave way to the German delegation, which argued that the wider the protection given to literature, the greater the advantage for the country giving the protection and that it was a matter of enlightenment and self-profit to extend copyright irrespective of reciprocity.

preparation of an international dictionary of technical terms of the trade and of an 'International Directory of the Book Trade'. The latter was published in 1912 and the *Vocabulaire Technique de l'Editeur en Sept Langues* in 1913. In its insistence on accurate bibliographical description the Association, as we have seen, had been ahead of continental practice, but again in May 1911 the Council urged upon members the importance of dating title-pages, and its deprecation of the post-dating of books published in the last month or two of the year, and reminded publishers of Guide Books, in particular, that since their value depends to a great extent upon their being up to date, the omission of the date of issue must be imputed to dishonesty.

THE OPERATION OF THE NET BOOK AGREEMENT

The Net Book Agreement emerged unscathed from the Book War with *The Times* in 1908 and in the understanding of its implications—and what it did not imply—the Association and the Associated Booksellers were generally in harmony. Its bearing on the Public Libraries was a matter of concern to both Associations. In May 1907 a Delegation from the Library Association, armed with a resolution from a conference of representatives of municipal and other non-commercial library authorities, had requested the Association for the concession of special terms of supply, as in the United States, and had been told that any concession must be subject to the assent of the Associated Booksellers. The Library Association then approached the Booksellers, but found itself unable to accept an offer made by them of a discount of 5% off net books combined with a limitation of the discount on non-net books to 25%. It also abandoned an alternative project for forming a trading association for the supply of libraries upon finding that it would be required to observe the conditions of the N.B.A. Again in October 1909 one of six resolutions passed by the Associated Booksellers at their annual meeting proposed that, in view of the successful way in which the net system was working, the time had come to consider the allowance of a fixed maximum discount to libraries, but the Council could only reply that whatever action might appear advisable should be taken by the booksellers themselves.

Of the other five resolutions four related to matters of general trade practice: premature remaindering, pre-publication subscription prices, allowances for copies held in stock by booksellers upon the publication of cheap editions, and the alleged need of booksellers for a higher rate of profit, to meet the capital risked, on books published at 10*s*. and over. On these proposals the Council resolved: that no books except novels should be remaindered within two years of first publication; that the offer of a pre-publication subscription price (to be raised on publication) was an indispensable means of publishing certain books; that an allowance for a book remaindered or reduced in price should be given only on copies which the bookseller could show that he had purchased within the preceding two years; and finally—not for the first or the last time—that the Association could not interfere in the terms granted by individual publishers to their customers. To the sixth resolution, which concerned the supply of school books to Education Authorities, we revert later in this chapter. In the acceptance of new kinds of retail outlets as entitled to be supplied at trade terms, these years brought the recognition in 1909 of the Café Cairo in Dublin (the first of what came to be known as 'other traders'), of the first bookstalls within universities and colleges (Manchester Municipal Technical School in 1910 and Imperial College in London in 1911) and of the Crown Agents for the Colonies in 1910.

THE CIRCULATING LIBRARIES

In the nineteenth-century march towards egalitarianism, education in state schools had gone hand in hand with self-improvement through books borrowed from the circulating libraries of Mechanics' Institutes and literary societies. During the period under review, and indeed for many years to come, the widespread subscription, or commercial circulating, libraries were still a powerful force in the distribution of the latest general literature. Credit for the invention is commonly given to Scotland, but the Edinburgh prototype of 1726 was quickly followed by others in Bristol, Bath, Cambridge, Norwich, Hull, Liverpool and Newcastle; and in London Mudie's, with its guinea subscription serviced from its large premises in New Oxford Street, became pre-eminent after the 1850s. It was indeed the circulating libraries which created the artificial vogue of the 'three-decker', the publication of new novels in three volumes at a guinea and a half for the set. The 'three-decker', which was at the height of its popularity in the 1860s and 1870s, continued until 1894 when, with the growth of the free Public Libraries and the cheap reprint and the ever-increasing popularity of the novel over other forms of literature, the libraries refused to take new novels in that form at its artificial price, and its place was taken by the one-volume novel issued at the non-net price of 6s. and until the 1914–18 war often purchasable for 4s. 6d. Nevertheless, the proprietors of the circulating libraries retained a powerful voice and in January and November 1908—during and after the settlement of the dispute with *The Times* Book Club—representatives of Mudie's, W. H.

Smith & Sons and Boots were received by the Council for the discussion of the time limit after which copies of non-net books, which meant in effect novels, might be sold at second-hand prices. Since an analysis by five of the leading publishers had shown that for the general run of novels 90% of the first year's sale took place in the first month after publication, the Council felt that it could safely accept three months, which the libraries favoured, rather than a longer period which would have been preferred by the Associated Booksellers.

In 1909 William De Morgan's new novel, *It never can happen again*,[1] was the subject of a boycott by some libraries, and in December of that year the newly formed Circulating Libraries Association addressed to publishers a letter about the dangers of 'obnoxious literature', with a request that its members should be given more time before publication to inspect books 'of a doubtful nature'. The Council after expressing sympathy with the purpose of the Libraries Association, proposed a joint meeting of representatives of the two Associations and the Society of Authors. After a false start by the Society, two of whose chosen representatives, Sir George Darwin and Edmund Gosse, declared themselves incompetent to discuss the subject, a meeting of the General Purposes Committee on 15 March 1910 was attended by Maurice Hewlett, Mrs Belloc Lowndes and George Bernard Shaw. Hewlett demanded that the unnecessary restrictions being placed upon the circulation of books by the libraries should be withdrawn and suggested that this end might be achieved if the Publishers Association acquired one of the libraries and ran it 'on common-sense lines'; and Bernard Shaw expressed the belief that if the matter were not dealt with, the libraries would attempt to dictate the size, price, etc., of books. No decision was reached, but the President, William Heinemann, undertook to bring up the matter at a General Meeting of the Association. Subsequently Heinemann had another meeting with Hewlett and A. D. Acland of W. H. Smith & Sons, and secured their agreement to a memorandum (which, tantalizingly, has not survived) 'on the Censorship embodying a proposed remedy for the existing unsatisfactory condition of affairs'.[2] At its April meeting the Council welcomed the proposal, but with 'the strong recommendation that the negotiations, and more especially the names of the censors, should be regarded as strictly confidential'. When the General Meeting was held in June there was no unanimity among the members and any attempt at a formal censorship came to its proper end.

[1] Heinemann initiated an attempt to price novels according to length by issuing this long novel in two volumes at 10*s*.   [2] Council Minutes, 21 April 1910.

THE SEVENPENNIES

The impact of cheap reprints upon the 'three-decker' novel has been mentioned. Just as in 1935 the first Penguins at 6d. were to signal the end of the several series of 3s. 6d. cloth-bound reprints, so in 1908-9 the profitability to author and publisher of the 6s. novel seemed to be threatened by the sevenpenny cloth-bound reprints of copyright titles. At its meeting in October 1908, the Council received a letter from Algernon Methuen, calling attention to the injury likely to result from the publication of sevenpenny editions, but the Council were of the opinion that the Association could not officially express any opinion upon the legitimate operations of some of its members and that those specially interested in the 6s. novel should take such unofficial action as they might consider desirable. Heinemann thereupon convened a meeting of the publishers of novels on 19 November and circulated a memorandum which might be sent to the Society of Authors and from which the following is an extract:

In our opinion, the issue of long copyright novels, bound in cloth, by distinguished writers (who are still producing 6s. novels) at sevenpence is certain to depreciate the value, not only of the individual author's forthcoming work, but also of the 6s. novel in general. A fair profit may be made by the reissue of an old novel at sevenpence in cloth form, but we are of the opinion that an author whose older works are available at sevenpence each in cloth will find, on publishing a new 6s. novel, the orders from librarians and booksellers substantially reduced, chiefly because once the public is thoroughly accustomed to the purchase of cloth-bound novels by living authors at sevenpence, it is certain to decline to pay 6s. only for the sake of novelty. If the custom increases, novelists whose books sell to a moderate extent are likely to be extinguished altogether, and popular authors to have their sales decreased, while new and untried authors will find it more difficult than at present to secure publishers.

We are of opinion that profit cannot be derived in the long run from both forms of publication. On the other hand the sixpenny paper novel has not in the past had the same evil effect on the sale of 6s. novels as the recent sevenpenny cloth book. The sixpenny novel has yielded considerable profit to authors and publishers and is therefore to be preferred in our opinion to the sevenpenny cloth book, in the case of authors who are still producing new work.

Of the twenty-four firms invited to the meeting nineteen were represented, and two resolutions were passed: that it was expedient that the publication of novels in cloth at 7d. and of standard copyright works of general literature at 1s. should be discouraged; and that a committee consisting of C. J. Longman, Frederick Macmillan, Reginald Smith, G. T. Hutchinson and Algernon Methuen should communicate with the

Society of Authors. A sub-committee of the Society, however, favoured the issue of sevenpennies after the lapse of two years from the date of original publication; and at the Annual General Meeting of the Association on 1 April 1909 Frederick Macmillan sprang a surprise by giving notice that his firm proposed to issue six sevenpennies. Nevertheless, at a second meeting on 19 April nineteen of the novel publishers reaffirmed the first resolution of 19 November, regretted the decision of the Authors' Society, and bound themselves, subject to the concurrence of the absentees, not to issue sevenpenny editions of any novels of which they owned the copyright and to do their best to dissuade authors from such editions of novels of which the copyright was in the author's hands. In the face of Macmillan's action and of the decision of the Authors' Society not all the absentees signed the resolution. When the signatories met again on 3 May Longman, in the chair, reported supporting evidence from the President of the Associated Booksellers. 'From the evidence of the last two years' wrote Henry Keay,

it is found that the 7d. editions of the best authors have made a very material difference in the sale of the 6s. editions, people refusing to give 4s. 6d. even for the new novels of those authors who have had 7d. editions of some of their writings...On the other hand, with works of authors like Marie Corelli, who refuse to allow 7d. editions, the sales keep up, the public realizing they cannot be purchased in the cheap form.

Although unanimity upon the resolutions passed at the previous meeting could not be achieved, the sixteen publishers present on 3 May resolved that it was inexpedient to issue at less than 1s. in cloth any 6s. novel within less than five years from the date of its first publication in book form and bound themselves, pending further negotiations with the Authors' Society, to implement that resolution until 1 July. These resolutions were conveyed to the Authors' Society and after a final meeting of the novel publishers on 16 June and with the consent of the Society, Longman reported the outcome to all the members of the Association, as follows:

The Authors' Society have been giving very careful consideration to the question of cheap editions of novels, and as a preliminary step have sent out a circular to their members asking their views on the subject. They have received 231 answers from writers of novels, and of these 203 are in favour of agreeing not to publish cheap editions within two years from first appearance, while only 28 take the opposite view. Thereupon the Authors' Society have decided to consider further the whole question of the price of novels.

In view of this most important announcement, it is hoped that publishers, without pledging themselves to any definite course of action, will use their best efforts to persuade authors to take no further steps towards the publication of novels of a less age than five years at less than 1s. until the Authors' Society have had time to consider the whole question thoroughly.

No more was achieved. But at the Council meeting in February 1910 the President called attention to the Annual Report of the Committee of Management of the Society of Authors, which deplored the decrease in the colonial sales of works of English authors and gave as a principal reason the issue of 7d. editions.

TAUCHNITZ EDITIONS

Travellers from this country to the Continent before, and after, the 1914–18 war will recall the welcome opportunity which they then had of purchasing the cheap reprints in the English language which the German firm of Tauchnitz[1] was licensed by the original publishers to produce for European sale only; and they will remember the temptation, to which perhaps they succumbed, to smuggle them into England on their return home. This lapse was sufficiently widespread to engage the pertinacious attention of the Council, for the Annual Report for 1909–10 records:

The Council have been in communication with the Commissioners of Customs to ensure a more careful examination of travellers' luggage, with a view to preventing the importation into the United Kingdom of books infringing British copyright. The Commissioners of Customs declared themselves unable to give the desired undertaking, and referred the Council to the President of the Board of Trade, who in his turn referred them to the First Lord of the Treasury. The First Lord of the Treasury in his turn referred the Council to the Chancellor of the Exchequer, who ultimately undertook that notices as to contraband and prohibited goods should in future be posted up on the steamers, and the attention of passengers directed to them.

The prevention of piratical importations of a more serious kind continued to depend upon registration of titles with the Customs authorities in this country and, for example, in the Australian States. The Copyright Act of 1911 required notice in writing to be given to the Customs and in this country the Customs remained unwilling to accept registration at Stationers' Hall; and when in November 1913 the General Post Office was requested to stop the entry by post of copies of an American reprint of Swinburne's *Atalanta in Calidon*, it replied that the entry had been properly allowed since the work did not appear in the Customs' list.

[1] The first volume of Baron Tauchnitz's 'Collection of British Authors' was issued in 1841 and when he died in 1895 some 3,000 titles had been published. From the start he recognized the rights of the authors, although there was then no international copyright protection, and also undertook not to export his editions to Britain or her Colonies.

THE EDUCATIONAL BOOKS COMMITTEE

The Educational Books Committee, which had been set up to deal specifically with problems arising from the 1902 Education Act, was constituted as a standing committee in April 1908. One of its earliest recommendations in that year was in support of the New Zealand bookselling trade, and the Council's ensuing action was the first example of its continuing policy to foster the growth of booksellers in British territories overseas. The New Zealand booksellers had reported that it was governmental policy that the regional Education Boards should purchase elementary school books direct from publishers in the U.K. The Council took up the matter with the Education Department in Wellington and although the Minister replied that the choice of channels of supply was to be made by the Boards, he affirmed his accord with the Association 'respecting the benefits that may be conferred on a community by the booksellers' and volunteered that the Departmental grants for the purchase of school books were sufficient to enable the Boards to purchase through the local booksellers, allowing them a reasonable profit. Of the thirteen Education Boards, to which the Council proceeded to write, five agreed to purchase through local booksellers, one to order the greater part of its supplies locally, and another to leave the purchase to parents and teachers, who were likely to buy locally.

I have mentioned that one of the six resolutions passed by the Associated Booksellers at their annual meeting in 1909 concerned the supply of school books to local Education Authorities. It advocated the institution of an agreement to limit the discount allowable on school (non-net) books to the Education Authorities, but on the recommendation of the Educational Books Committee the suggestion was rejected by the Council on the grounds that fewer school books would be purchased, that the Association would be brought into collision with the L.E.A.s, and that the school supply contractors would represent themselves as being coerced by the Association.[1]

The Committee tried unsuccessfully in 1908 and 1909 to persuade the London County Council to withdraw its ban on visits by publishers' representatives to the schools within its authority and to withdraw its standing order prohibiting teachers from asking for inspection copies of books. Much time was also spent in 1908 and succeeding years in discussion with the National Union of Teachers, which expected the publishers'

[1] It appears that at this time the school contractors had a domestic agreement not to supply at a greater discount than 37½%, a discount larger than they received from the publishers.

exhibitions of books to contribute, both educationally and financially, to the success of its annual conference, but was reluctant to provide adequate facilities.

FIRST CONTACT WITH TRADE UNIONS

The year 1913 brought the Association into contact, for the first time, with a trade union. A dispute arose between the bookbinding firm of James Burn & Co. and the National Union of Bookbinders and Machine Rulers, and the Union warned publishers in September that it had instructed its members to refuse to touch the work of any publisher who placed orders with Messrs Burn during the continuance of the dispute. The President, J. H. Blackwood,[1] called a special General Meeting of the Association on the 12th, at which it was agreed that members should reply that they intended to maintain absolute freedom of action and that the Association, which had been asked to receive a deputation from the union, should decline to do so until the ultimatum had been withdrawn. It was subsequently withdrawn and on 25 November the Officers received a deputation, but the Council found that the dispute dealt only with technicalities of the binding trade and decided to take no action; and again in April 1914 it declined a request from the London Trades Council that the Association should endeavour to bring about a meeting between the two sides.

THE PUBLISHERS' CIRCLE

It remains to be recorded that in the summer of 1908 the Council gave its avuncular blessing to the formation of a luncheon club, which had been suggested at the Annual General Meeting by Arthur Spurgeon of Cassell's. The Publishers' Circle, which still continues to provide a forum for the discussion of new ideas, was founded in July, with Arthur Waugh of Chapman and Hall as its first President and A. D. Power as honorary secretary.

[1] of William Blackwood and Sons, known particularly for *Blackwood's Magazine*, of which the publication has continued in unbroken succession since its first publication in 1817.

4

1914-1919
World War I

LEIPZIG EXHIBITION

'I would like to urge the members of the Publishers Association to induce as many of their employees as possible to take Leipzig in on their holidays this year. That they will come back improved in knowledge and usefulness there is no question, and I believe that they will find the trip very agreeable.' Many members of the trade were soon to make extended trips to the Continent which they did not find agreeable. The quotation is from a report on the International Exhibition of Books and Graphic Arts, which Heinemann presented to the Council on 21 May 1914.[1] This Exhibition, the most comprehensive one of its kind yet held, was opened in Leipzig on 6 May by the King of Saxony, but the British Pavilion, which was organized by the Board of Trade and, like several others, was not ready, had three days' respite because the British Ambassador ordered it to remain closed on account of the death of the Duke of Argyll. The main British exhibit was constructed as a library, with subdued lighting, into which visitors escaped from the glare of the more showy displays elsewhere. 'The display of English literature' wrote Heinemann, 'is a complete revelation to most visitors of the activities of English publishers, as isolated books only find their way ordinarily to the shelves of Continental booksellers, and not always the most attractive ones.' The Exhibition was prematurely closed by the outbreak of war on 4 August.[2]

[1] *Members' Circular* I, no. 19 (May 1914), 142.
[2] In November 1926 the *Members' Circular* (VI no. 21, 230–2) reprints an interesting memorandum on the pre-war concentration of the German book industry in Leipzig and on Leipzig's loss of its pre-eminence to Munich during and immediately after the war. 'Before the war it was accepted as a fact that every tenth individual in Leipzig earned his bread through the production and sale of books, and the number of workpeople of all kinds engaged in the different branches of this industry exceeded that of any other in the city. This is not now the fact. The war period gave a great fillip to the metal industry, which now has pride of place in importance compared to the book industry'.

52 1914–1919: World War I

WORLD WAR I

The Council held a special meeting in September, at which it received a suggestion from the Publishers' Circle that a Committee should be set up to deal with all trade matters affecting members of the Association and especially problems arising from the war. The Trade Committee was constituted in the following month.[1] But contrary to expectations the war during its first year had little effect upon publishers generally, apart from some decline in sales, and did not increase the work of the Association. In the Annual Report on the year 1914–15 J. H. Blackwood, who had succeeded Sir Frederick Macmillan as President in 1913, referred to the war in two connections only. It had been feared that some firms might be misled into thinking that the war with Germany cancelled the international copyright of German authors. This the Council thought to be a matter upon which its opinion ought to be publicly stated and the President wrote a letter to *The Times* of 7 October 1914, warning those whom it might concern that the rights of individual Germans in their own works must be respected, inasmuch as there had been no abrogation of the conditions of the Berne Convention. The second matter referred to concerned book reviewing by the press. A report had spread that reviews were to be suspended during the war, but a general inquiry about the intentions of newspapers had produced almost wholly satisfactory replies, and the hopes which they held out had not been disappointed.

PAPER SUPPLIES AND PRICES

By the end of 1915 the loss of merchant ships and the heavy demands upon the diminishing tonnage began to necessitate a reduction in the importation of paper-making materials, esparto grass from North Africa and pulp from North America and Scandinavia. At the first of three meetings of the Council in February 1916 Reginald Smith, who had been elected President again in the previous March, reported that he had been informed by the Board of Trade that the Government proposed to appoint a Royal Commission on Paper,[2] whose duty would be to fix the principle on which supplies were to be distributed among users and to

---

[1] W. M. Meredith (Constable & Co.) was Chairman until he became President of the Association in March 1917, when he was succeeded by Humphrey Milford (Oxford University Press).

[2] The original title appears to have been The Royal Commission on the Importation of Paper. The first Chairman was the Rt Hon. Sir Thomas Whittaker, a Liberal M.P., Chairman of the United Kingdom Temperance and General Provident Institution and active in the temperance movement.

superintend its application, with power to control the prices charged; and he reported also that he had been asked to submit the names of four members of the Association of whom one would be chosen to serve on the Commission. The Council nominated four members and fearing that the comparatively small needs of books were in danger of being drowned by the larger voices of other users of paper, they agreed to ask the President of the Board of Trade to receive a deputation. At a second meeting in the same month Smith reported that Sir Frederick Macmillan was the chosen representative and he read the case which he had put to the Board of Trade for the maintenance of educational and scientific books and of cheap reprints for war-time relaxation and for the maintenance of exports in the face of American competition: the first of many arguments by successive Presidents in two wars for the recognition of books as something more than articles of commerce. But the request for a deputation was rejected and the case, together with a similar one from the Society of Authors, was referred to the Royal Commission. By April 1916 publishers were becoming alarmed at the prices being charged by paper-makers and in the hope that the Royal Commission might be induced to regulate them, Heinemann undertook to prepare a report for the Trade Committee, and the Secretary began to keep a register of paper and other book-making materials which members were willing to sell. Heinemann seems to have done his utmost to investigate prices without bias. He found that an ordinary printing paper which could have been bought for $2\frac{1}{8}d.$ per lb. in 1915 had gone up to $5d.$ and he found some evidence of higher prices being asked. He recognized that a rise in price followed inevitably from the limitation of imports of raw materials to two-thirds of 1914 supplies and from the appointment of a Royal Commission which, being primarily concerned with the importation of raw materials, was unable or unwilling to control prices and yet by its very existence frightened paper-makers into precautionary increases. He regretted also that no steps had been taken to collect waste paper and called attention to an expired German patent for the remaking of paper, which was in use in Germany. Heinemann concluded:

in view of the fact that the Paper Commission has shown no sign of trying to alleviate, by careful husbanding of the national resources, the hardships inflicted through a national necessity, the conclusion is forced upon me that the Commission is mischievous as far as the publishing business is concerned. Certainly it has failed in distributing evenly the burden of the War on all those concerned in the making and the use of paper.

Finally, he recommended that the Board of Trade should ensure that users of paper should not get less or, by offering higher prices, more than two-thirds by weight of what they used in 1914; that the profits of the manufacturers or importers should be limited to their 1914 level; that waste paper and cardboard should be made use of; and that the Royal Commission should inquire into the existing comparative rises in costs and materials. The report, which was communicated anonymously to members in June 1916,[1] was referred to Macmillan as the Association's representative on the Royal Commission.

Whether or not Heinemann's investigation and proposals contributed to a more equitable control of quotas,[2] there were no more general complaints until the end of 1917. Regulations limiting the distribution of catalogues and prospectuses were imposed in March 1917, but as distribution from 'trader' to 'trader', i.e. publisher to bookseller, was still permitted and as the Royal Commission allowed that a schoolmaster was a 'trader' in school books, it was taken to follow that a chemist, doctor or theologian would also be 'traders' in books bearing on their occupations. The effect of the regulations was therefore not serious. The processing of waste paper was under way by May 1917, when an Order was made fixing the maximum prices at which it might be sold; but a protest to the Royal Commission that publishers were getting a raw deal from control of prices on the waste paper which they had to sell and no control on those of the finished paper which they had to buy brought only a reply from the Secretary of the Commission that the question of fixing prices for new paper had not been overlooked and that waste paper was mainly used for making paper for Government purposes, a statement which must have been received with some ironical pleasure. November 1917 brought further restrictions on the circulation of catalogues and prospectuses to the public, limiting the amount of paper used to one-third of the weight used in the corresponding period of the previous year, but distribution to booksellers and to schoolmasters continued to be exempt.

In December 1917 it was announced that the quantity of finished paper and raw materials to be imported during 1918 for trade use, excluding Government requirements, would be reduced from 540,000 tons to 390,000; and the consternation of publishers provoked a spirited defence of the Royal Commission from Sir Frederick Macmillan. 'The shortage of shipping' he wrote[3]

---

[1] *Members' Circular* I, no. 31, 232 ff.
[2] Sir Stanley Unwin, whose firm, George Allen & Unwin, came into existence on the day war was declared and therefore had no pre-war basis for a quota, records having on occasions to pay 1s. 7d. per lb. for paper inferior to what he could have bought for $2\frac{1}{4}d$. before the war. (*The Truth about a Publisher* (1960), p. 141.)
[3] *Members' Circular* II, no. 11, 114.

due partly to the losses by submarine and partly to the demands made upon tonnage for the transport of American troops and material, has obliged the War Cabinet to take powers to cut down to the lowest possible figure the import of everything that is not Food, Munitions or Men. Neither wood-pulp nor esparto comes under this category...I can assure you that the Chairman of the Royal Commission on Paper did his best for us with the War Cabinet and I have reason to think that but for his efforts the measures taken might have been more drastic...It has been suggested in some quarters that publishers of books might ask for special consideration on the ground that what they produce is of national importance, but I feel certain that such a claim to preferential treatment could not be supported with any chance of success.

Nevertheless the Council, noting that the Royal Commission, by granting licences to import strawboard for the binding of educational books, had already acknowledged the right of such books to a measure of priority, appointed a committee of three[1] to meet the Chairman of the Commission and to propose that the control of the allocation of all paper should be taken over by the Government and that a system of priority certificates should be inaugurated, in order to provide publishers with more paper at reasonable prices. At a meeting in February the Chairman[2] said that the Commission intended to control supplies and prices more than in the past and that under new regulations shortly to be issued practically no 'free' paper was to be expected; and the possibility of priority certificates for educational books and the standardization of book-papers were discussed. Although a plan for standardization was put to the Paper Controller, nothing came of it, but in March the proposal for priority certificates, for medical and educational books only, was adopted and the Association was invited and agreed to undertake the duty of reporting upon the applications. An advisory committee of three, consisting of C. F. Clay,[3] Gerald Duckworth and Henry Scheurmier was appointed and was recognized by the Controller as acting for his department.[4] Although he limited the definition of an educational book to one which 'was expected to be in the hands of every scholar in the class', he did concede that 'books needed for class use must be produced at any cost'. In practice a priority certificate ensured that a publisher would get the paper to which he was entitled, but at first it gave him no right to any extra tonnage above his quota.

[1] Humphrey Milford, Heinemann and Henry Scheurmier (Thomas Nelson & Sons).
[2] Sir Thomas Whittaker had been succeeded by Sir Henry Birchenough, Chairman of the British South Africa Company and appointed by the Government to be Chairman of several other advisory committees on commerce and industry during the war.
[3] Cambridge University Press.
[4] When the control of paper ceased on 30 April 1919, the Committee had held eighteen meetings and had considered 121 applications for certificates.

In March 1918 the Royal Commission was superseded by a new Advisory Council of the Controller of Paper, composed of representatives of all the trades interested in the manufacture, distribution, and consumption of paper in its various forms. In addition to Sir Frederick Macmillan, who remained a member, the Association was invited to nominate two representatives, and the President and the Vice-President, Meredith and Humphrey Milford, were chosen. But the paper available from reduced imports was insufficient to meet all needs and at the end of May the President sought the support of the President of the Board of Education, H. A. L. Fisher, for the allocation of more shipping tonnage for educational books; and for the information of the Paper Controller he asked members confidentially to inform him of the total quantity of paper used by them on all their publications in 1917. The Board of Education replied that it had already made representations on the need for more paper for educational books and could do no more, but July brought the welcome news from the Paper Controller that he had made arrangements to import an additional 5,000 tons of pulp per month from Sweden, which would be distributed as an extra ration proportionately among all licence holders between August and the end of the year, and that holders of Priority Certificates who failed after every possible endeavour to obtain the paper from manufacturers in this country would be granted special import licences. The returns made by members showed that in 1917 they had used no more than 11,588 tons on the production of books, as an addition to which in the current year educational books were estimated to require 2,000 tons. The claims of books in terms of tonnage were small indeed—less than 14,000 tons out of the 390,000 tons which had been announced as available for all trade users of paper during 1918— and as the autumn advanced towards the Armistice, publishers' worries about supplies diminished.

It was otherwise with prices. The Controller was prepared on occasions to intervene to secure the reduction of prices above what he regarded as the market level and in December 1918 he issued the first of a promised series of monthly reports on reasonable prices, listing an ordinary book-paper (which Heinemann had priced at $5d.$ in 1916) at $9\frac{1}{2}d.-11d.$ per lb. The easing of supplies after the Armistice had brought some fall in prices. But it was temporary and after the precipitate ending of the Paper Control on 30 April 1919 publishers were before long to pay as much as $1s.\ 9d.$ per lb. for an inferior substance which they would have rejected at $2\frac{1}{2}d.$ before the war.

SHORTAGE OF OTHER MATERIALS

Although the scarcity of paper was the first concern of the Association and its Trade Committee as the German submarines tightened the blockade, fears of shortages in other materials required for book production and concern at mounting prices occurred from time to time. By April 1917 supplies of strawboard, both British and imported, were much reduced and the price had risen from £4. 10s. to £35 per ton. Following a suggestion from the Publishers' Circle that paper-covered books might become generally unavoidable, the Council issued a warning through the press that the public must be prepared to accept books, particularly those for schools, in paper or limp cloth bindings. In spite of a report that the Food Controller intended to withdraw the licences to use flour and starch in the finishing of bookcloth, an assurance came from the Royal Commission on Wheat Supplies in May 1918 that no such restriction was intended. At the end of 1916 the use of copper, otherwise than for munitions, was prohibited and the process-engravers were able to secure supplies for block-making and electrotyping only against the release of old blocks and electros which publishers no longer required. Throughout 1917 there was an increasing demand for lead, particularly for the making of shrapnel bullets, and in December the Minister of Munitions, Winston Churchill, invited representatives of the printing, type-founding and publishing trades to a conference to secure the release of superfluous type metal and obsolete stereotype plates. A National Committee for the Release of Metals was formed, including three representatives of the Association,[1] who in March 1918 requested members to release plates and again in the following month, after the Minister had expressed disappointment at the first result, urged them to a further effort.

DEPLETED STAFFS

As the war advanced, publishers' staffs became increasingly composed of older men, women, and boys below the age of eighteen. A report in January 1917 that all men up to fifty-six years of age were to be withdrawn from non-essential trade caused the Council to submit to the Director of National Service, Neville Chamberlain, arguments for the classification of publishing as a trade of national importance.[2] But the only concession

[1] Edward Arnold, C. F. Clay and Sir Frederick Macmillan.
[2] Although no statistics of book exports were then available, it was thought that the amount was likely to be so small in relation to the total exports of the U.K. as to weaken the case and accordingly any reference to exports was deliberately deleted except the dependence upon them of schools in the Dominions and Colonies.

granted was permission to employ National Service volunteers awaiting call-up or transfer to an essential trade, and in March 1918, when the withdrawal even of men discharged from the Forces was threatened, further representations were made to Sir Auckland Geddes, Director-General of National Service. In April, with a new National Service Act in operation, he was pressed again for some concessions and in the following month he gave an assurance that employers of men covered by the new Act, but placed in a low medical category, could apply to the tribunals for their retention on occupational grounds; and, with a directive to the tribunals in view, he asked for comparative figures of the number of employees engaged then and before the war in publishing and of the numbers concerned with various kinds of book. A limited assurance was also received about the employment of discharged soldiers: that a firm dealing in books of national importance would be permitted to engage and retain them. In June Meredith submitted the figures for which he had been asked. Omitting firms who were merely British agents for foreign or colonial houses or were mainly printers and stationers or had been forced out of business by war conditions, he based his returns on sixty-two firms out of the total membership of eighty-two. There were 1,559 men of military age and over and 1,522 women and boys under age, as compared with 2,891 and 658 respectively in June 1914. Of the 1,559 men 429 were over fifty-one years of age, 90 had been discharged from the army, and many others had been registered as unfit or in low medical grades. Their occupations ranged from 101 working directors and partners at the top to 685 warehousemen on the floor. Although Meredith found it as impossible then as it would be now to classify the staffs according to the kind and the supposed importance of the books with which they were dealing, he gave the following percentages of the books in the current lists of the sixty-two firms: scientific, theological and educational, including works of reference, 65%; fiction and children's books, 17%; war books and general literature, 18%. In July, with the war almost at its end, there came a reply from the Ministry of National Service that it was proposed to include book-publishing in a new list of certified occupations with the effect that men in the medical grades I, II and III would be protected at age 45, 35 and 25 respectively. Great satisfaction was expressed at the result of the President's representations.

RISING COSTS AND PRICES

Increasing costs of manufacture and their impact upon the long-established level of book prices presented no less of a problem than the progressive reduction of supplies and labour. By October 1915 bookbinders felt compelled to increase their charges and a Cost of Production Committee was appointed by the Council. The first action of the Committee was to receive a deputation from the Associated Booksellers, who were generally in favour of meeting higher costs by the extension of net prices to categories of book still being published at prices subject to discount, but were not unanimous about the inclusion of novels, school books and rewards.[1] The Booksellers undertook to consult all their members and the Council advised publishers to make no general increase in prices in the meantime, although recognizing that the binding of new supplies of books published at 1s. or less, in which the cost of binding was a predominant part of the cost, might necessitate immediate increases.[2] When in January 1916 no further report had come from the Booksellers, members became impatient and the Council appointed a committee to consider net prices for novels and to consult the circulating libraries, and it referred the application of net prices for school books to the Educational Books Committee. That Committee proposed that publishers should meet their increased costs on educational books by reducing the trade allowance on net prices to twopence in the shilling and, believing that booksellers were generally reducing the discount given by them, by supplying non-net books to the trade at 'scrip'.[3] The circulating libraries were in favour of net prices for novels. The report of the Associated Booksellers, when it came in February, showed that its general members had supported a resolution recommending the fixing of net prices for all books, with a minimum trade allowance of 25%, but that net prices for school books had been opposed by the school traders who, treating school books as 'loss leaders' in their tendering for the supply of stationery, were believed to be allowing discounts of 37½–40% to the new Local Education Authorities. The Associated Booksellers could do no more than recommend that the sale of all books at net prices should be tried out regionally.

Within the Association the attitude of the Council, no less in war than in peace, was that the raising of prices or the limitation of trade terms were not matters on which corporate action could or should be taken, and even the issue of a public manifesto on the inevitability of increasing prices was abandoned in April 1916. In February 1917 the Associated Booksellers

[1] see p. 71, n. 2.
[2] By the end of the war the 7d. and 1s. cloth bound reprints of copyright titles had disappeared and the price of the 'World's Classics' and 'Everyman's Library' had risen by degrees from 1s. to 2s. a volume.
[3] 'Scrip' (a contraction of 'subscription price'): 'a trade price 25% below the retail price'— O.E.D. This meaning of the word appears to have been confined to the book trade.

repeated its resolution of a year before, reminding publishers also that more books were being successfully issued at net prices, that there had been local successes in limiting the discount allowed on non-net books, and that a trade allowance of twopence in the shilling was insufficient to cover booksellers' overhead expenses, which had risen to 18–20% of sales, and protesting against the action of three publishers in particular—Macmillan, Bell, and Arnold—who by reducing their trade terms rather than raising their published prices were seeking to force an increase in actual prices paid for non-net school books by the education authorities. But the only issue which continued to engage the attention of the Council was the possibility of getting the publishers of novels to adopt net prices. Although five firms found themselves unable to agree to a general resolution, the 6s. novel had generally become a 7s. or 7s. 6d. net book by the end of the war.

IMPORTS FROM U.S.A.

The war-time limitation of imports brought in March 1917 a prohibition upon the import of books other than in single copies by post, which imperilled the securing of copyright by American authors in Great Britain and thereby throughout the countries of the Copyright Union. Even though the securing of copyright did not require manufacture of an edition in this country, effective publication required a supply of copies sufficient to meet 'the reasonable demands of the public'. Once again the opinion of MacGillivray was sought, but he could only reply that the effect of the prohibition was that publication in the U.K. would not serve to secure copyright throughout the countries of the Union unless an edition were set up in the U.K., and he advised that recourse should be had by American publishers to publication in Newfoundland (which, as we have seen, as a self-governing colony had adopted the Copyright Act of 1911), with the reminder that publication in Canada would not be effective since that Dominion[1] had still not adopted the 1911 Act. In June 1918, to meet this problem and, in reverse, the difficulty of securing in war conditions copyright in British works by manufacture in the United States, the Board of Trade proposed an Order in Council instituting a moratorium in the U.K. until six months after the end of the war, on condition that the United States would grant a similar moratorium upon the necessity to register and manufacture there.

[1] It was in fact in 1917 that 'Dominion status' for the self-governing parts of the British Empire was finally recognized.

COPYRIGHT IN GERMAN BOOKS

The Council continued its policy, which it had voiced in October 1914, of discountenancing the translation of books for which the war made it impossible to make terms with the authors; and the same policy was adopted in Germany towards British books. In October 1915 a suggestion was made to the Comptroller of the Patent Office that just as it was possible to obtain licences to use German patents during the war, so the licensing of translations of German books and the collection of royalties should be undertaken by some Government Department; and the Comptroller expressed his willingness to promote the necessary legislation. But the Board of Trade was at first unwilling to move unless sufficient examples were forthcoming of books of which English translations were required in the national interest, and the Council was able to quote only a book on explosives. In June 1916, however, the Comptroller reported that the Law Officers of the Crown had considered the problems of copyright in books published in enemy countries both before and during the war. They advised that the copyright of books published before the war was not destroyed by its outbreak and that, if necessary, they could be vested in the Public Trustee under the existing Trading with the Enemy Act, without any breach of the Berne Convention, but that the unlikelihood of their needing republication in this country made such action unnecessary. As regards books published during the war, they held that its outbreak had suspended the operation of the Convention and that such books were unprotected. An Act was accordingly to be introduced, to vest their copyright in the Public Trustee and to empower him to grant licences for their publication or translation and to charge royalties. The Act was passed, the ultimate destination of the copyrights and the royalties being left for determination at the end of the war. At the request of the Comptroller the President of the Association attended the hearing of the first application for a licence; and the Association undertook to advise its members to limit applications to books required in the public interest rather than for private entertainment. The Association continued to uphold the principles of the Berne Convention and when in October 1917 H.M. Stationery Office used the Act to apply for a licence to translate two German works on optical instruments published before 1914, the Council protested against an unnecessary breach of the Convention. When the Peace Treaty was signed in 1919, it granted immunity to German reprints and translations which, in defiance of the rights of the British copyright owner, had been printed and published between 4 August 1914 and 28 June 1919 and gave the German publisher the right to continue to sell them within German territory until 27 June 1920.

NATIONAL BOOK FORTNIGHTS AND CHRISTMAS CATALOGUES

Although problems of supplies and labour took up a large part of the time of the war-time Presidents and Councils, attention was given also to sales promotion both at home and abroad. In January 1915, depressed by the effect of the war upon sales, and particularly upon the smaller booksellers, and stimulated by reports of the success of a co-operative publicity bureau of American publishers, the Trade Committee began to give thought to the issue of a Christmas Catalogue and in July to a scheme put forward by Arthur Spurgeon of Cassell's for a National Book Fortnight; and the two ideas were developed in conjunction. Arrangements were made for the compilation and production of the Christmas Catalogue, and for the collection of paid advertisements from publishers to be undertaken by *The Publishers' Circular*; and that journal and *The Bookseller*[1] were appointed joint agents for its sale to booksellers and other trade purchasers. Of the 1915 issue 48,000 copies were printed and distributed in mid-November; and the account showed a small surplus which was divided among the advertisers. The National Book Fortnight, the first nation-wide effort of the Association to enlarge the circle of book-buyers and to get the press to support book-buying as a habit, was held from 22 November to 4 December. The principal London, Scottish and provincial daily papers each devoted two pages, in which were printed special articles contributed free by H. G. Wells, Arnold Bennett and other authors, classified guides to recent books, and publishers' advertisements; and the selected papers and others supported the purpose of the Fortnight in their news columns and leading articles. Window-banners and show-cards were exhibited by over 500 booksellers, who also made extensive use of the Christmas Catalogue. The Fortnight was pronounced by the Associated Booksellers to have been a success, but with less enthusiasm for the Catalogue, and it was agreed by both parts of the trade that it was only by co-operative action that an effective effort could be made through the press to gain new book-buyers. The Fortnight, supported by an improved Christmas Catalogue, was repeated in 1916 and was publicized in a wider selection of newspapers, including particularly those in provincial cities containing munition works, where money was plentiful.

[1] For an account of the two trade papers, see p. 106.

EXPORT CLEARING HOUSE

In the export field also the Association was active, helping members to maintain sales to allied and neutral countries, so far as war-time conditions allowed, and to be ready to increase them when the war should end. When France introduced import licensing in 1917, some relaxation of the formalities for British books was secured; in July 1916, as a counter to German propaganda in neutral countries, the Council circulated an annotated list of Continental periodicals likely to review British books; and the attempt of some American publishers, beginning in 1915, to detach Australia and New Zealand from the British publishers' market was opposed.

More significantly, in June 1916 the Trade Committee recommended that a co-operative export clearing house should be set up in London, to undertake the collection and dispatch of British books to Continental booksellers 'on sale or return' and to prepare and distribute suitable advertising material—in short, to match the facilities available to German publishers through their co-operative agency in Leipzig. An Export Agency Committee,[1] working at high speed, produced a detailed scheme, which was circulated to members in July, backed by support in principle from the Society of Authors and by provisional encouragement from the bodies representing the weekly newspapers and periodicals, the technical and trade journals, and the music publishers. It was stated that such was the efficiency of the Börsenverein in Leipzig that booksellers in Scandinavia, Holland, Italy, Russia, and even in Spain and South America, had become accustomed to order British books through Leipzig, and that the war-time substitution of German books and the resulting propagation of German 'Kultur' were causing alarm to the Foreign Office. The scheme proposed the establishment of a comparable British organization, to which a subsidy from the Government might be expected, and made detailed recommendations for its representative management, for the payment by publishers of a 10% commission on sales made through it, to cover its expenses, and for the subscription by publishers of £25,000 capital, of which £5,000 would be raised in cash and £20,000 by the deferment of payment for books sold during the first eighteen months.[2] To the scheme was appended a letter from the Foreign Office saying that a Government grant—originally put at £1,000 but increased after discussion to £5,000—

[1] Consisting of G. S. Williams (Chairman), Heinemann, Humphrey Milford, and W. Longman.
[2] The scheme is set out in detail in *Members' Circular* I, no. 32 (July 1916).

was under consideration. While a decision was awaited, action was deferred also on a suggestion from the Department of Commercial Intelligence at the Board of Trade that, in order to escape the duty on bindings recently introduced by the Italian Government, paper-covered editions should be produced for sale to that and other continental countries. It was January 1917 before the decision came, and it was that the Chancellor of the Exchequer did not consider that a case had been made for a state subsidy. Believing that the Treasury had decided irrevocably that the required grant had to be viewed as no more than a subsidy to a trade, and therefore refused as a dangerous precedent, and believing that without Government help the capital and the staff could not be found, the Council saw no alternative but to abandon the scheme.

The Export Agency Committee, nevertheless, was kept in being and at a meeting in May 1917 it resolved that it was the duty of the Association to formulate a fresh scheme for the better distribution of British books abroad and that, while any such scheme must involve the sending of books 'on sale or return', the establishment of a central clearing house would mean that publishers would not be required to open more than one 'on sale' account. When the Committee met again on 7 June it learned that the Royal Society had set up a committee of three to consider the promotion of British books and periodicals abroad and that the chairman, Dr H. K. Anderson,[1] was of the opinion that any suggestion of a Government grant towards an undertaking which would be judged to be purely commercial would still be fatal and that accordingly, if Government support were to be obtained, the request for it had better be made by the Royal Society. The Committee, for its part, considered that, while publishers could be expected to support an export agency only if it had a reasonable probability of becoming self-supporting, Government support was indispensable for its launching. However, on 20 June the President received a letter from the Director of Information, John Buchan,[2] saying that the Prime Minister had authorized him to appoint a Departmental Committee to consider the circulation of British books abroad. 'The Germans' he wrote, 'have largely controlled in the past the educational and intellectual world of Europe, owing to the special prominence given to their publications.' The terms of reference of the Committee would be to frame a scheme, including a clearing house, for the

---

[1] Master of Gonville and Caius College, Cambridge, later Sir Hugh Anderson; Chairman of the Syndics of the Cambridge University Press from 1918 to 1928.

[2] The author of *The Thirty-nine Steps* and other novels of great popularity; later, as Lord Tweedsmuir, Governor-General of Canada; in 1907 had joined Thomas Nelson & Sons as literary adviser, to superintend their sevenpenny editions of 'The Best Literature'.

sale of British books in allied and neutral Continental countries, with provisions for long credit and the return of unsold copies, which would be acceptable to local booksellers; and he would ask the Treasury for a grant. He proposed that the Committee should consist of: one representative each of the Royal Society and the Royal Society of Literature; one representative of the Universities; one representative of the publishers of technical periodicals and one of the publishers of popular magazines; and three representatives of the Publishers Association. The Council appointed the President (W. M. Meredith), C. J. Longman, and G. S. Williams as its representatives; and those appointed by the other bodies were: Dr H. K. Anderson (Royal Society), Sir Henry Newbolt (Royal Society of Literature), Professor W. MacNeile Dixon (Universities), G. E. Chadwyck-Healey (technical periodicals), W. Grierson (magazines).

While the Departmental Committee was sitting, the Association forwarded suggestions to it, in particular offers from the Copenhagen Booksellers' Association of better facilities for the display of British books and from the Swedish Booksellers' Association to found a special importing house, although the latter offer had to be followed by a warning that the person to be put in charge of it was reputed to be a notorious German propagandist. In December members of the Association were advised that the Committee's report might be expected early in 1918, but it was not until November that the Report could be circulated to members and a General Meeting called to consider it. The Report proposed the establishment of an elaborate British Books Distribution Organization, to be financed substantially by the publishers of books, periodicals, and magazines, but with no indication of the extent of Government assistance. At the General Meeting on 27 November a letter was read from the Weekly Newspaper and Periodical Proprietors' Association expressing sympathy with the proposed organization, subject to further knowledge of its constitution and to financial support from the Government and contingent upon the members of the Publishers Association financing it to a greater extent than would be expected of themselves. Undeterred, the fifty firms represented at the General Meeting resolved that it was in the interest of the nation and the trade that the recommendations of the Departmental Committee should be brought into effect and that the members of the Association should support them financially; and a new Export Agency Committee was appointed to confer with the Association of Trade and Technical Journals (which had not been represented on the Departmental Committee), and to form with them a joint deputation to the Government.

When the Committee reported in June 1919, it produced a depressing document. The Weekly Newspaper Association had said that it could not promise a financial contribution without a more definite plan of the organization than the Departmental Committee's Report had contained and without an estimate of the capital required, but had offered no assistance in the preparation of any plan or estimate. The Association of Trade and Technical Journals had said that its members had not heard officially of the Report and that few had expressed willingness to consider it. The Music Publishers' Association and the Fine Art Guild, who had also been consulted, had offered no support. Perhaps the emphasis in the Departmental Committee's Report had seemed to be too much on books and too little on other publications. Trying to drive in one firm peg somewhere and believing that it was pointless to work out in detail the methods to be employed, and the probable cost involved, in carrying out the recommendations of the Departmental Committee without first ascertaining the probable amount of financial support within its own Association, the Export Agency Committee sent a circular to the members. It produced from twenty-five firms a promise of £7,050 in capital and £11,650 in credit. With such small support within the Association, and little or no encouragement without, the Committee saw no alternative but to recommend that the Association take no further action unless approached by the Government with a definite offer of financial support; and the Council agreed. No doubt money was tight and energy at a low ebb after the worry and overwork of the war, but it is hard to resist the conclusion that the Council, which included a high proportion of seniors who had borne the burden of the Association's work for many years, might have shown more determination. An opportunity for a co-operative drive for exports was lost, and although the Council was active in encouraging the enterprise of foreign booksellers and agents, the continental trade was left to the initiative of individual publishers, of whom Sir Stanley Unwin must be acknowledged as a pioneer, and of the London exporters, who were spurred to fresh efforts by the breakdown of the co-operative scheme. It must also be remembered that £5,000 then was a much more significant sum than it is now.

THREATENED LEGISLATION

Of the Government's war-time concern, or disregard, for books it remains only to record two instances of threatened legislation. In 1916 the Board of Trade began to give consideration to measures for securing the position of certain branches of British industry after the war and a Report of an

Advisory Committee[1] included recommendations, based on representations from the printing trade, that the United Kingdom Copyright Law should be brought into line with that of the United States (i.e. that manufacture in this country should be a condition of copyright) and that there should be an import duty of $33\frac{1}{3}\%$ on all printed matter. Although the Council protested that the opinion of the Association should be heard before any action came to be taken, the Council was divided on the issues involved: some members feeling that the only way of getting the United States to change its copyright law was to give it a dose of its own medicine and that, with the probability that British costs would have risen steeply by the end of war, protection might be needed against a flood of cheap, pirated reprints of British books from the United States; other members being firmly anti-protectionist in principle and opposed to any infringement of the principles of the Berne–Berlin Conference. When the war ended, the Report of the Board of Trade's committee remained in its pigeon-hole. The other legislative threat, against which the Association and the Society of Authors jointly demanded to be heard, came in 1918 from the recommendation of a Select Committee of Parliament for a tax on luxuries, including books, but the tax was not introduced.

PRINTERS AND BINDERS: WAREHOUSING CHARGES

The war-time increase in costs and the limitation of supplies compelled printers and binders to look into their methods of charging and, in particular, to institute separate rates for warehousing publishers' stocks, which had previously been generally included in their charges for printing and binding. In March 1916 the Master Bookbinders' Association announced in its *Circular* that from 1 May its members would charge for the storage of unbound sheets at certain rates which were quoted; and protracted negotiations began between committees of the two Associations. While a majority of the Trade Committee of the Publishers Association, insensitive to the bargaining position of smaller publishers, would have preferred that charges for storage, as for binding, should be subject to competitive tender between binder and customer and should be at lower rates for larger than for smaller customers, the Council recognized that members could not be left to fight individual battles and that in some cases the binders had a just claim to make a charge for warehousing stock, both unbound and bound. The concern of the Council was then to ensure that rental charges would be accompanied by corresponding decreases in binding prices, and to try to secure a free initial period after the delivery

[1] *Parliamentary Paper* Cmd 8181.

of sheets to the binder from the printer and a scale of charges related not only to the number but also to the format of the books in store and related to the frequency and to the size of binding orders for each book. The negotiations were complicated also by questions arising from the binders' liability to protect the stock which they held from fire, damp, and vermin, and their right to a lien on it against unpaid accounts; and it was not until February 1917 that a settlement was reached. The terms which members were recommended to accept provided, in brief, that with effect from 1 July 1916 there should be different rates of charge for small, medium, and large books; that no charge should be made during the first nine months; and that binding prices inclusive of storage would be subject to an appropriate rebate.

In July 1917 the printers began to follow the example of the binders, by instituting charges for the storage of unbound sheets and for the rental of standing type. After the Master Printers' Federation had declined to appoint a committee to meet a committee of the Association, negotiations were opened with R. A. Austen-Leigh, of Spottiswoode, Ballantyne & Co., who undertook to convene a group of book-printers; and detailed proposals were put to him for mitigating what seemed to be inequities in the charges proposed, and for rationalization comparable with that achieved with the binders. In the charges for warehousing the Trade Committee sought differentiation between small cheap books and large expensive ones on the scale, with the free initial period, to which the binders had agreed; and in the rental for standing type it sought a free period of twelve months—to give the publisher time to decide whether to distribute the type or to have moulds made or to keep it standing permanently—and differentiation between type set by hand and by Monotype, in which the value of the metal was not the same. The printers replied that conditions in the printing and binding trades were not comparable, that they could not extend the free period of three months which was given, in effect, by their system of rendering accounts, and that any distinction between the two kinds of type-metal would make difficulties for printers who continued to set by hand. They offered concessions in the graduation of warehousing charges according to the number of pages in a book, but not according to its format. The difficulties of the Committee were further increased by the fact that it was dealing with an unofficial group of printers, and the five whom it met declined to give the names of the others whom they represented. Negotiations were broken off in June 1918 and the Council could only recommend members to give their business to firms which were moderate in their demands.

TRADE UNIONS

In another field at this time the Association had helpful co-operation from the Master Printers' Federation, and indeed became an affiliated member of the Federation for a short period. In 1918 the National Union of Printing and Paper Workers, having secured recognition in the printing trade, began to extend its activities to the staffs of the wholesale booksellers, W. H. Smith & Sons and Simpkin Marshall & Co., of whom the latter was a member of the Publishers Association; and at a special General Meeting in June it was agreed that the Association should become affiliated to the Federation, in order to have the benefit of its experience in dealing with trade disputes. By 1 August the Union had threatened to call out the employees of certain firms, and at another special General Meeting later in the month the Association resolved, while disclaiming recognition of the union, to appoint a committee to advise members who might be attacked. When the Committee met in September to consider whether a standard rate of wages for warehouse staffs could be drawn up, it learned that four firms, including Messrs Simpkin Marshall, had been given notice by some of their employees and had had a meeting with the Union officials to discuss the minimum wages demanded. At later meetings the Committee rejected, as economically impossible, a schedule of wages proposed by the Union and, although a resolution in favour of recognizing the Union was carried by the casting vote of the chairman, it was subsequently rescinded. By March 1919 doubts had arisen in the minds of the leaders of the Association whether it was constitutionally competent to deal with a trade union and at a further special General Meeting that month it was agreed that a distinct Book Publishers' Employment Circle should be constituted, whose membership would not necessarily be confined to members of the Association and for whose acts the Association would not be legally responsible. With an abatement of the Union's activity other than in the enrolment of members, the Employment Circle was not required to meet until October 1920 when it offered to mediate in a fresh dispute involving Messrs Simpkin Marshall. It was not until 1925 that acute differences were to arise leading, as we shall see, to a three months' strike in which the whole of the London book-distributing trade was involved.

CONSTITUTION: COUNCIL AND GROUPS

In other ways also the Association gave thought to its constitution and organization. At a Special General Meeting in February 1916 the Rules were altered, to provide that at least three of the thirteen members of the Council should retire annually, and at the Annual General Meeting in 1917 the Rules were again amended, to include among the duties of the Treasurer responsibility for the preparation of the *Members' Circular* reporting more fully the transactions of the Council. The realization that the increasing work of the Association imposed a heavy burden upon the Council and that its decisions should be more broadly based upon the general opinion of members led the Council in 1917 to give effect to a suggestion which had come from the Publishers' Circle that members should be formed into groups according to the class of publications in which they were interested. Eight Groups were formed: I General literature other than Fiction; II Fiction; III Educational, secondary and elementary; IV Juveniles; V Cheap Popular libraries; VI Medical, Scientific and Technical; VII Bibles and Prayer Books; VIII Maps. Each Group was to appoint its own chairman and secretary; and its resolutions were to have validity as acts of the Association only when adopted by the Council. This Group system, with some variation and with differing degrees of usefulness, has lasted until now; and Group III in particular had a record of continuous activity in dealing with the corporate problems peculiar to the publication of school books.

Group III held its first meeting on 2 May 1917 and elected C. J. Longman as its first chairman. Among the precedents which it inherited from the former Educational Books Committee, which ceased to exist upon the formation of the Group, was a decision in 1916 to recognize the West Riding Education Authority as entitled to buy books for school use at trade terms, upon condition that it opened a bookstore; and by 1920 the Council had accepted recommendations from the Group that five other Education Authorities be similarly recognized as booksellers. The Group had also to consider its policy towards the fixing of net prices for school books. Although in 1916 the Educational Books Committee had recommended against this practice, some publishers continued to have recourse to it as a means of increasing their receipts, as costs rose, without increasing the published price. The practice brought protests from Education Authorities that they should not be asked to pay the same price for a quantity of a book as they would for a single copy; and as post-war costs became stabilized the fixing of non-net prices again became general and accepted as inevitable by the booksellers.[1]

[1] In September 1927 the Group unanimously approved a resolution of a committee which had been appointed to examine the proportion of school books still being published at net prices, 'that, in future, all books intended primarily for class use, both elementary and secondary, should be published at subject prices [i.e. not at net prices]'.

The only other Groups to meet before the end of the war were Group II (Fiction) and Group IV (Juveniles). Group II, with Arthur Waugh as chairman, defined the London area within which orders were delivered free and recommended minimum trade terms[1] for both net and non-net novels, but the efforts of a majority to reach a general undertaking to issue all novels at net prices were frustrated by absentees and, as we have seen, were unsuccessful. Group IV, with Arthur Spurgeon in the chair, received a reminder from the Council after its first meeting in May 1917 that its resolutions would not be acts of the Association unless confirmed by the Council, but nevertheless a number of members put their signatures to an undertaking to issue their 'Reward'[2] books, as far as possible, at net prices and to limit the terms allowed to booksellers for both Juveniles and Rewards. A move out of the area of price maintenance into the dangerous one of price-fixing was made in November 1917, when the Group agreed—but not unanimously—that the price of Children's Annuals published that year at 3s. 6d. net should be made 5s. net for the 1918 issues and that Annuals published at prices below 3s. 6d. net should be increased by at least 6d.

AUTHORS AND LIBEL

Before the war the Council had instructed a committee to revise the Association's model forms of publication agreements between author and publisher, to bring them into line with the 1911 Copyright Act. The Committee had the co-operation of the Society of Authors and the revised forms were printed and circulated to members in February 1917, but doubts arose about the legal enforceability of the clause which required the author to give a guarantee that his work 'contains nothing of a libellous or scandalous character'. Counsel's opinion[3] emphasized the unsatisfactory state of the law of libel and focused the attention of the Association and the Society of Authors upon the dangers of unconscious libel. After much discussion between the two bodies it was at last agreed in February 1920 to substitute a new clause providing, in particular, for the publisher to share equally with the author damages and costs arising 'where any matter contained in the said work shall be held to constitute a libel upon a person to whom it shall appear the author did not intend to refer'. But the law continued, and continues to this day, to leave author

[1] The Associated Booksellers at their Annual Meeting in June 1917 passed a resolution requesting publishers so to price their net books as to be able to allow the trade $33\frac{1}{3}$% on orders for two or more copies.
[2] 'Rewards' were books for children of a kind suitable for prizes in Sunday Schools.
[3] By D. M. Hogg, later Viscount Hailsham, and F. T. Barrington Ward.

and publisher open to attack from some unknown person who saw in it the opportunity of fortuitous damages.

END OF AN ERA

The war years exacted their toll upon the leaders of the Association. Reginald Smith died in December 1916, during his Presidency, and in October 1917 John Murray, overworked and anxious about his son at the front,[1] was ordered by his doctor to reduce the time which he could give to the Association's affairs. Indeed, the old order, which had done so much for the Association in its early years and carried its burdens in the war, was about to give place to the new. Before the end of the 1920s death was to take Edward Bell, William Heinemann,[2] John Lane, Sir Algernon Methuen, and Murray; C. J. Longman would have retired; and of the old guard only Sir Frederick Macmillan would remain. But in the meantime, to set against the anxieties of war, the members of the Association had some occasions for rejoicing. They took pride in the award of a V.C. to 2nd-Lieut. Frederick Palmer, a publisher and the brother of a publisher, in the conferment of a Knighthood on Arthur Spurgeon in 1918, in the election of a member of the trade, Sir Horace Marshall, as Lord Mayor of London for 1918-19, and in the knowledge that the United States Ambassador at the Court of St James, Walter H. Page, had been a publisher.[3] In the general rejoicings at the end of the war there was a felicitous exchange of congratulations between the Association and its sister associations in France, Belgium, Italy and the other allied countries.

[1] Lieut.-Colonel John Murray, D.S.O. (John Murray V).
[2] Heinemann bequeathed to the Association the sum of £500, to be retained 'as a reserve for the Council in view of any emergency where the interests of British publishers may be threatened'.
[3] of Doubleday, Page, Inc., author of *A Publisher's Confession* (Heinemann, new ed. 1924).

5

1919-1927
Reconstruction and strikes

As war-time conditions quickly receded during the early 1920s, the normality to which the publishing trade began to return was normality with a difference. New problems and new opportunities arose and to the younger leaders who came forward, both in the old firms and in new ones,[1] the traditional policy of the Association seemed too conservative and slow-moving.

COUNCIL UNDER CRITICISM

As we have already seen, John Lane in 1907 had criticized the older statesmen of the Association for lack of initiative and for their limited conception of its functions; it had been pressure from the unofficial Publishers' Circle which had led to the formation of Groups within the Association; and in January 1920 the Circle was prodding the Council again, this time to tackle the problem of increased rail charges and the consequent demands for help from booksellers. Since the establishment of the Net Book Agreement in 1900, the primary objective of the Association had been to defend, rather than to initiate new ideas or to promote co-operation between publishers and booksellers. Because defence of the Net Book Agreement included defence of booksellers from themselves, the Council seemed to see the retail half of the trade not as partners but as juniors needing to be kept in their place with severity and on occasions even with acrimony. Not only with booksellers but also with authors and librarians there was a need for greater co-operation in the cause of books and this need was soon to bring into being the Society of Bookmen and the National Book Council, the forerunner of the present National Book League. Within the Association itself the importance of individual liberty of action had been paramount, but in July 1920 *The Bookseller* was reporting that the younger minds were thinking that the usefulness of the

[1] Jonathan Cape (with his partner G. Wren Howard), Geoffrey Faber and Victor Gollancz established their firms in the 1920s.

Association was much hampered by the understanding that resolutions of the whole body were not necessarily binding upon all its members. But publishers were to remain individualists and many years later a member of the Council could be heard complaining with one breath that it was senseless to pass only a recommendation, which members would not be bound to follow, and with his next insisting that his firm could not possibly limit its freedom of action by acceptance of a binding resolution.

Of the Council's difficulties in reconciling the counter-claims of freedom and co-operation the critics were no doubt insufficiently aware, but we cannot dismiss the charge that it was insensitive to new ideas and movements and too ready to minute that action would be 'not within the Council's jurisdiction'. In the later 1920s, to take one example, British typography and printing were revitalized by the work of Stanley Morison and the Monotype Corporation and in the early years of the decade the younger men in publishing were already alive to what was happening,[1] and were beginning to give it expression in the production of their books. Yet in 1922 the Council rejected a proposal from the British Institute of Industrial Art for a typographical exhibition, to be held at the Victoria and Albert Museum, on the ground that 'an exhibition illustrating artistic British book production would not be of service to them'.

COUNCIL ELECTIONS

The Council's ability to adapt itself in the late 1920s to new conceptions of its role will be the subject of the next chapter, but in the meantime it was once more the Publishers' Circle which gave expression to the feelings of the younger men by putting forward in July 1920 proposals for regular retirement of members of the Council. Its recommendation, based on the rule of the French Cercle de la Librairie, for a three-years term of service was accompanied by the following note:

The working of the rule which obliges members of the Association to vote for not less than 13 candidates for the Council under penalty of the voting paper being invalidated, makes it easy for firms other than those of historical standing and the highest rank to secure representation for one year but almost inevitably results in their losing it in the year following. As a consequence of this, such firms have frequently declined nomination... Opinions with regard to important questions affecting publishing do not always find expression adequate either to their intrinsic importance or to the extent to which they are held among members of the Association.

---

[1] The monthly 'Book-Production Notes' which B. H. Newdigate began to contribute to *The London Mercury* in May 1920 deserve remembrance.

The Council immediately agreed to appoint four of its number, under Humphrey Milford's chairmanship, to meet two representatives of the Circle. This Joint Committee at its first meeting accepted the need for a method of election to the Council which would secure greater freedom of choice, and at a second meeting put forward recommendations for an amendment of the constitution. The new rules, which were approved in general meeting in January 1921, provided that the affairs of the Association should be conducted by the three officers (President, Vice-President and Treasurer) elected annually, and by a Council of twelve elected (as firms) for three years, four members retiring each year and being eligible for re-election.[1] Ineligibility after two consecutive terms was to wait for a later amendment, and no action was taken upon two other suggestions from the Circle, that the Association should employ a full-time Secretary and that its functions should not exclude the passing of resolutions binding upon its members.

EXPORTS: OPPORTUNITIES AND OBSTACLES

Although the Council, disheartened by the refusal of a Government grant and by the lack of sufficient financial support within the trade itself, had felt obliged to abandon the scheme for a centralized distributing house in London with branches abroad, it spent much time on other ways of assisting British publishers to seize the new opportunities open to them on the Continent. But the difficulties in the early 1920s were great; devastation in France and Belgium and economic depression in Denmark, Norway and Finland with consequent depreciation in their currencies made it difficult for them to buy foreign books. 'The high price of the pound sterling' wrote the President, Humphrey Milford, in the Annual Report on 1920-1, 'in almost all foreign countries has been and still is a great drawback to the development of export business, it being almost impossible at present, for this reason, to sell books in Central Europe and in the Near East.' In January 1921 the manager of the shop which W. H. Smith & Sons had opened in Brussels was putting before the Council the problems which he was facing with the Belgian franc at 59 to the pound, as compared with the pre-war rate of 24; in 1923 economic conditions and the rate of exchange in Denmark caused the abandonment of a projected exhibition of British books; and throughout Central Europe the impoverished universities were being starved of British books by the abnormal rates of exchange. In Germany international trade in books was in chaos. The demands of the Allies for the payment of reparations caused

[1] In 1930 the rule was amended to provide that individuals, not firms, should be elected.

the German Government to devalue the mark and the wholesale printing of paper money brought the rate to 74 to the pound in May 1921 and to 310 in November and thereafter in galloping ruin to 50,000 in December 1922 and to 21,000 million in November 1923.

If the Council missed opportunities of developing corporate action in the cause of British books in Europe, it did investigate and pass on to the members of the Association many suggestions from foreign booksellers' associations, from anglophile institutions and from the Department of Overseas Trade, which would help them individually to increase their business. When in 1925 the International Institute for Intellectual Co-operation was preparing for the League of Nations a report on postwar difficulties in the international circulation of books, the Council did no more than pass on its request for suggestions to members of the Association; and it failed at first to realize the possibilities of the Florence Book Fairs. But, on the other hand, the Export Agency Committee, which remained in being after the collapse of the scheme for a co-operative clearing house, had spent much time in 1919 in investigating and suggesting improvements in a subsequent offer by the London exporters, William Dawson & Sons, to institute a scheme of their own for the supply of Continental booksellers 'on sale or return'. At the same time the Committee tried also to promote the issue, by the proprietors of *The Publishers' Circular*, of a classified monthly list of new books, but the high cost of producing it and the difficulties of foreign exchange caused its abandonment. Success, however, attended representations which the Council made through the G.P.O. to the Postal Union Congress, and from 1 January 1922 an increase in the limits of weight for foreign post made it possible to dispatch a book not exceeding $6\frac{1}{2}$ lb. at the reduced rate applicable to printed papers.

FLORENCE FAIRS

International Book Fairs were held in Florence in 1922 and 1925 and provided a market-place for booksellers and book-buyers and a meeting point for publishers from many countries of a kind with which the current Frankfurt Fairs now make us annually familiar. When in February 1922 the organizing committee warned the Association, through the British Institute in Florence, of the support forthcoming from Germany and urged it to see that British books were well represented, the Council's first reply was that it could not move in the matter because the cost was out of proportion to the probable sales and because the rate of exchange in Italy was unfavourable to Britain, but favourable to Germany. When the Italian Government offered free space for a British exhibit and the payment of all transportation charges, the Council appointed a committee to organize the section and about 1,000 books were hastily dispatched, but when the Fair closed at the end of July, the Department of Overseas Trade reported that the French and German sections had stolen the limelight. It was otherwise with the second Fair in 1925. H.M. Ambassador in Rome, Sir Ronald Graham, urged British exhibitors not to fail on this occasion, and the support of the Association and of leading publishers was canvassed by Mrs G. M. Trevelyan, the honorary secretary of the British Italian League, who was also successful in securing a grant of £500 from the Department of Overseas Trade. A committee consisting of two members of the Council, a representative of the Department and Mrs Trevelyan organized the collection and dispatch of an exhibit of some 10,000 volumes from seventy-four firms, and G. B. Bowes, the Cambridge bookseller, spared a member of his staff for three months to take charge in Florence. The Fair was opened by the King of Italy on 3 May and closed on 5 July. The value of the British exhibit in the promotion of international understanding was warmly acknowledged by Sir Ronald Graham, and when the President of the Association, G. S. Williams, made his annual report on the year he was able to record that the British section had been, according to independent sources, the most effective of any and, with a touch of smugness, 'a revelation to the Italians'. A third Florence Fair was held in 1928, but the Department of Overseas Trade was unwilling to provide a grant and there was no British exhibit.

IMPORTS FROM GERMANY

The internal problems of the German publishing trade resulting from the war do not fall within the scope of this narrative, but one impact of the attempt of the Allies to exact reparations deserves mention. Under the German Reparation (Recovery) Act of 1921 a British publisher, or bookseller, importing copies of German books was required to pay 26% of the invoiced price to the British Commissioners of Customs and was liable consequently to pay only 74% to the exporting German publisher, who was then to recover the balance from his own Government. But in November 1923 the German Government announced that it would no longer reimburse its exporters and the latter began to insist on British importers paying the full invoiced value and to reinforce their demand by supplying through Vienna. However impossible the monetary payment of reparations may have been—and the German Government made it worthless by its devaluation of the mark—this levy on German exports seemed to offer a practicable contribution. But in February 1924 by agreement between the two Governments the rate of the levy was reduced from 26% to 5% and the German Government undertook again to reimburse its exporters and to make it a punishable offence to charge any part of the levy to the British importer. In other markets of the world German scientific text-books continued to be dominant, and not least in Japan, where in January 1925 the British Commercial Counsellor reported that German rather than English was the foreign language used in the sciences, but nevertheless suggested useful ways in which British publishers could increase their business.[1]

REVIVAL OF THE INTERNATIONAL CONGRESS

Feelings of bitterness towards Germany in the early post-war years coloured the attitude of the Council towards the re-establishment of the International Congress of Publishers. In 1916 after a visit by William Heinemann to the Permanent Office in Berne the Council noted that its activities had been conspicuous by their absence, the Secretary having been temporarily called up for military service, and in the following year the Association reduced its subscription to the Congress from 2,000 to 1,000 francs. When in January 1919 T. Fisher Unwin, motivated perhaps by his nephew, Stanley Unwin, raised the question of the Association's attitude towards a revival of the Permanent Office, the Council decided that it was not then prepared to consider meetings between its representatives and representatives of Germany and the former Austria–Hungary and it favoured a suggestion from the Cercle de la Librairie for a meeting in Paris of inter-allied publishers only. In the following month also a letter from Victor Ranschburg, the Hungarian President of the last Congress, pleading the cause of his country, received an uncompromising reply from the Council, that Hungary must take 'the full consequences of defeat' in the war into which the Central Powers had plunged the world and that he himself should set himself to 'prove by deeds the radical change of mind to which you lay claim'. Although the proposed conference in Paris did not take place, G. S. Williams, the Association's representative on the Congress executive, was guided by Louis Hachette, the principal French representative, and it was at the suggestion of the French that in June 1920 a circular, signed by the French, Belgian and British representatives, was sent to the allied and neutral member Associations, proposing that the Permanent Office should be closed and the Archives

[1] After the Japanese earthquake in 1923 a grant of £25,000 was made by the U.K. Government to the Tokyo Library Committee for the purchase of British books.

deposited with the Confédération Helvétique in Berne. The replies showed an equal division for and against closure and the Council continued to follow Williams's advice that since personal intercourse among the full membership of the Congress would be impossible for an indefinite period, there was nothing for any permanent organization to do. By November 1921 Ranschburg, feeling that he personally stood in the way of a new start, resigned the Presidency and Van Stockum of The Hague, as Vice-President, put forward new proposals to keep the organization alive. But the French objected to the management being in his hands, on the ground that he was '*trop Bosche*', and the Council supported its French ally in advocating formal closure. That did not take place and a provisional committee, self-appointed, came into being. But the Association maintained its negative attitude until March 1928 when, at the instigation of Louis Hachette, the Council agreed to send two delegates, W. G. Taylor (of J. M. Dent and Sons) and Stanley Unwin, to a conference in Berne in 1929 to consider the re-establishment of the organization under new leaders and the holding of a new Congress. That a restart was possible was due to Ove Tryde of Denmark and Hans Lichtenhahn of Switzerland, who had kept the organization alive during the war, and to Gustav Gili of Spain, who had persistently advocated its importance throughout the 1920s.

CUSTOMS AND COPYRIGHT

The immediate post-war years brought some developments in international copyright and in international restrictions on free trade. French restrictions on the import of books ended in June 1919, but a movement in Canada in the same year to end the duty on British publications came to nothing. Both in Canada and in the United States orders were made by the Customs authorities requiring imported books to be marked on the title-page with the country of origin, to which the Association objected as a disfigurement made superfluous by the presence of the printer's imprint, then usually placed at the end of the book. The Canadian requirement was suspended for a time and when it came into force on 1 November 1922 it appeared to be satisfied by a printing imprint on the last page of a book. The requirement of the U.S. Treasury in the same year was qualified by a concession that the back of the title-page was equivalent to the front and, although in 1924 the Customs authorities became disposed to accept a printing imprint on the final page, the printing of the words 'Printed in Great Britain' on the verso of title-pages was in general use and had come to stay.

By the terms of the Peace Treaties of 1919 with Austria and Czechoslovakia Austria herself, which had not been a party to the Berne Convention but had had a separate copyright treaty with the United Kingdom, and the new countries which had constituted parts of the Austrian Empire were required to join the Convention. By 1923 the larger Indian States had agreed to adopt the provisions of the Indian Copyright Act of 1914 and so to give protection within their territories to works published in the British Empire and to receive protection for their own works in British India. It was thought by the Government of India that the need to give protection in India to books published outside the Empire or to protect indigenous publishers outside British India would not arise. In 1925 reports from the British Mission in Moscow aroused vain hope that a new decree recognizing personal copyright in works originating in the Soviet Republics might be followed by agreements between the Soviet Union and other countries; and although in the same year a joint committee of the Association, the Music Publishers' Association, and the Society of Authors spent time studying the draft of a new Chinese copyright law, it was decided that 'it would not avail us much as British subjects to interest ourselves in the perfecting of the Act as a legislative measure unless we could obtain from the Chinese Government definite and satisfactory assurances with regard to the adequate administration of the law when passed'.[1] Chinese piracy continued to flourish; and threats came also from the Near East. After the end of British rule Egypt did not join the Berne Convention and the announcement in 1924 that Arabic would become the medium of instruction in its medical schools soon brought from Egyptian publishers expressions of their intention to translate British text-books without permission or payment and from the Ministry of Education in Cairo a curt refusal to intervene.

In the United States the influence of the Typesetters' Union prevailed and manufacture there continued to be a condition of copyright. In 1923 a new Bill, to amend the law of 1909, was introduced and a long discussion of its purport in the correspondence columns of *The Times Literary Supplement* was opened by a letter from Thring, the Secretary of the Society of Authors.[2] The Bill gave automatic protection to foreign authors belonging to any country of the Copyright Union without registration or other formalities—and without local manufacture. But, whether by an error of drafting or not, the manufacturing clause of the old Act was not repealed, so that although a foreign author would obtain American copyright, his

[1] *Members' Circular* VI, no. 2 (February 1925), 18.
[2] *T.L.S.* 22 March 1923.

books, if in the English language, could not be imported and he could enjoy its fruits only by an American printing. The sponsors of the Bill withdrew it, the Act of 1909 continued, and the United States remained unable to join the Copyright Union. As, however, reprinting by photo-lithography, so called, came into use, American publishers sought increasingly to establish copyright in books by British authors by the use of that process and British publishers became aware that, whereas a sale of sheets from their editions would make some contribution to their typesetting costs, the negotiation of a reprint might bring none. Although the Council was made aware of this problem in 1926, it was not until 1932 that, under pressure from Stanley Unwin, a two-way resolution was agreed with the American National Association of Book Publishers, recommending a financial arrangement for the photographic reproduction of books composed on one side of the Atlantic and subsequently printed on the other.[1] The Association was also to take up in the 1930s the allied need for the protection of typographical design by copyright.[2]

In Canada, as we have seen,[3] new copyright legislation did not immediately follow the Berne Convention of 1911 and when a new Act was passed in 1921 it was thought to be contrary to the provisions of the Convention and to be detrimental to the interests of copyright-owners in the U.K. The Act included compulsory licensing clauses, which in effect provided that if the owner of the copyright did not print his book in Canada or grant a licence for that purpose, any Canadian publisher could apply for a grant of a compulsory licence to print on payment of a royalty and, in order to make the compulsory licensing system effective, Section 27 prohibited the importation of copies printed outside Canada. But the Canadian Government which passed the Act, being apparently advised that it would be a breach of the Berne Convention to prohibit the importation of books lawfully printed in the U.K. or in any country of the Copyright Union, exempted such books from Section 27. After the Act was passed, but before it came into force, there was a general election and the new Government, being less protectionist, passed an Amending Act of 1923, which provided that the compulsory licensing clauses also should not apply to the works of citizens of the U.K. or of any Union country. In 1926 a new and improved Act was passed, which as regards British copyright-owners not resident in Canada gave copyright whether an edition was printed in Canada or whether copies of the British edition

---

[1] In *The Truth about a Publisher*, pp. 383–4, Unwin criticizes the recommended basis for the contribution.
[2] see p. 123.   [3] see p. 41.

were imported and enabled the owner also to protect a Canadian publisher who printed an edition or imported sheets of the British edition by authorizing the Canadian publisher to prevent, as an infringement of the Canadian copyright, any importation of the book by another person.

REGISTRATION AT STATIONERS' HALL

At home the occasional difficulty in establishing legal proof of copyright gave rise to consideration of a system of registration. Under the Act of 1842 registration at Stationers' Hall had been compulsory and on 1 January 1924 the Stationers' Company opened a new Register in which publishers might voluntarily enter titles and dates of first publication, upon payment of a fee and the deposit of a copy of the book. In the following March the Master of the Company convened a meeting of interested parties, and the usefulness of the Register and of its extension to cover unpublished works, in which under the 1911 Act copyright existed no less than in published works, was advocated in particular by Sir Frederic Kenyon of the British Museum and Thring for the Society of Authors. But C. F. Clay, as President of the Association, saw dangers in voluntary registration, which might lead to an assumption that any work not recorded in the Register was public property, and he was supported by the representatives of the music publishers and the professional photographers. Nevertheless increasing use was made of the Register and more than 1,000 entries were made in 1925; and in January 1927 a second conference was called by the Stationers to consider whether the Register could be given authorized status. MacGillivray brought to the meeting a draft Bill to enforce registration as under the 1842 Act; and Sir Frederic Kenyon suggested that to facilitate entry the statutory free copies for the British Museum and the other copyright libraries should be delivered via Stationers' Hall.[1] But the possibility of statutory enforcement was not pursued, and although most of those present saw the advantages of compulsory registration as evidence of ownership, they saw no way of avoiding a conflict with the Berne Convention which had abolished all such formalities.

[1] It is interesting to note that when a quarter of a century later the *British National Bibliography* was established, the British Museum was unwilling for its deposit copies to come to it through the *B.N.B.* and, since publishers declined to provide an additional copy for the compilation of the *B.N.B.*, insisted that particulars required for the entries should be taken down on Museum premises.

COPYRIGHT LIBRARIES: LIMITED EDITIONS AND IMPORTATIONS

No doubt the provision of yet another free copy in the cause of copyright was not welcomed and publishers continued to kick against the pricks of the six libraries of deposit. George Moore's *Avowals* and *Héloise and Abelard* were issued in limited editions 'printed for private circulation and sold only to subscribers' and his publisher, Werner Laurie, claimed that this was not publication, which for the purposes of the Act meant 'the issue of copies of the works to the public', and that accordingly the British Museum had no right to demand a copy. But in July 1920 F. D. Mackinnon, K.C.,[1] gave a contrary opinion, from which the following extract may be quoted:

I am clear that 'publication by subscription' in itself does not fall outside the definition of an 'issue of copies to the public'. Even if Dr. Johnson had been able to print a list of subscribers ('Sir, I have two very cogent reasons for not printing any list; one, that I have lost all the names, the other, that I have spent all the money'), and if every copy of his Shakespeare was subscribed for before publication, I think it would have been 'issued to the public'. It can make no difference whether copies of a book are offered for sale, to anyone who likes to buy, before the actual copies are available for delivery or after.[2]

There was also resistance by publishers to the deposit of free copies of a book which they imported from the United States as agents of the American publisher, and in July 1920 there was quoted against them an opinion of the Law Officers of the Crown that a publisher was liable for any book bearing the name of his firm, even though he was acting as agent for another. The Association again consulted Mackinnon and received from him the unwelcome opinion that an agent is nonetheless the publisher in the U.K. and that it is immaterial whether his name appears or not.[3] It became, however, the practice of the libraries not to demand American books unless they bore the name of a publisher, an agent, or an agency address in Britain, and a letter of 11 July 1922 from the British Museum confirmed this limitation.[4] A question also arose in 1920 about the rights of the National Library of Wales under its Board of Trade Regulation of 1912, which laid it down that a book which was not a Welsh book need not be supplied if the number of copies of the published edition did not exceed 300.[5] Here Mackinnon's opinion was that the

[1] later Lord Justice Mackinnon.
[2] *Members' Circular* III, no. 17 (July 1920), 159 ff.
[3] *ibid.* III, no. 19 (November 1920), 188 ff.   [4] *ibid.* IV, no. 16 (July 1922), 148.
[5] There were other limitations on books published in limited editions at prices in excess of £5, recapitulated in *Members' Circular* V, no. 11 (December 1923), 92.

published edition was effectively the number imported, rather than the number printed by the American publisher, and that an agent importing less than 300 copies in one or more lots was not liable to provide a copy for Wales. Nevertheless, when in 1924 the Board of Trade issued new Regulations for this Library, the existing clauses about limited editions published originally in the U.K. were accompanied by a new clause, to which the Council had previously agreed through Sir Frederick Macmillan, its representative on the Council of the Library: that the books to be delivered should not include 'any foreign book imported into England of which the number allotted to Great Britain does not exceed 100 copies'.

TRINITY COLLEGE, DUBLIN

Upon the establishment of the Irish Free State in 1922 some publishers acted upon the assumption that the number of the copyright libraries had been *ipso facto* reduced by one, but a letter to the Colonial Office brought a reply on 18 April, on behalf of Mr Secretary Churchill, that the supply of free copies to the Library of Trinity College, Dublin, would continue in force until altered by legislation affirming the new constitution. So it remained, in defiance of logic and in the cause of Anglo-Irish unity in scholarship, with neither His Majesty's Government relieving British publishers of the provision of free copies for Trinity College nor the Government of the Irish Free State relieving Irish publishers of their obligations to supply free copies of their publications to the five libraries outside Ireland. When in 1926 the Irish Industrial and Commercial Property Bill was published it was found that certain of its provisions relating to copyright would have required the Free State to repudiate the Berne Convention, but the Bill was amended to give effect to alterations which the Association put forward on the advice of MacGillivray.

BROADCASTING

Early in 1923 publishers began to be aware of wireless broadcasting and of the British Broadcasting Company of Magnet House, Kingsway, as a potential user of copyright material in a new way or on a new scale. In April both at the Annual General Meeting and at the monthly meeting of the Publishers' Circle immediate consideration of the problems was urged and the Association was invited by the Society of Authors to support discussions which it was already having with the Broadcasting Company. The Council then set up its first Broadcasting Committee,[1] which pro-

[1] C. F. Clay (President), G. S. Williams, H. Scheurmier, G. Duckworth, G. G. Harrap.

ceeded to confer with Major I. H. Beith[1] and Thring, respectively Chairman and Secretary of the Society of Authors, and to negotiate with the Broadcasting Company. At the first meeting in May Mr J. C. W. Reith (as he then was), the General Manager of the Company, said that although his legal advisers were of opinion that broadcasting did not necessarily come within the Copyright Act of 1911, nevertheless he was prepared, without prejudice, to enter into negotiations for the use of literary property and to mention the titles of books and the author's name, and indeed the publisher's name unless that were prohibited, as advertising, by the Company's contract with the G.P.O. But at that time the Company's interest in literary material did not extend much beyond the broadcasting of stories for children. By midsummer an understanding had been amicably reached and the Association and the Authors' Society undertook to recommend to their members, during a trial period of one year, the acceptance of a fee of £1. 1s. per 1,000 words of prose passages and a minimum fee of £1. 1s. for poems not exceeding 200 lines in length or £2. 2s. per act or canto or division exceeding 200 lines. These fees, with a reduction for poems of less than 100 lines, were formalized in July 1924 in an agreement between the Society of Authors and the Broadcasting Company and were subsequently renewed annually. Broadcasting of extracts from books was not found to be detrimental to author or publisher and the Association did not join the Music Publishers' Association in giving evidence before the committee which the Postmaster General set up in 1925 to advise him upon the renewal of his licence to the Broadcasting Company. The sale of sheet music was suffering and the Company would be even less able to pay compensating fees for the use of copyright music if the Committee were to recommend that the public should be able to enjoy broadcasting for less than 10s. at which the annual licence fee then stood. The Music Publishers' Association stressed also the coming increase in the range of broadcasting and the impossibility of collecting fees for continental reception. 'This difficulty' wrote their Secretary, 'seems to point to the necessity of asking the Postmaster General to so limit the power of the stations erected by the British Broadcasting Company that no such direct reception abroad can be possible.'

Although our Association did not share this alarm, in 1928 it began to be apprehensive of the growing activity of the British Broadcasting Corporation (as it had then become) as itself a publisher of educational books and pamphlets. Legal advice was taken as to whether this was

[1] widely known under his pen name, Ian Hay.

permitted by the Charter, and the opinion of Counsel[1] was that it undoubtedly was permitted. Similar alarm was aroused in the newspaper world by the publication of *The Listener* and on 11 January 1929 representatives of the Association[2] joined the Newspaper and Periodical Proprietors' Associations in a deputation to the Prime Minister. Baldwin suggested that the deputation should endeavour to meet the B.B.C. in conference, failing which he would give his considered opinion. That afternoon Sir John Reith met the leaders of the three delegations and within a few days there was a meeting between the full deputation and the Governors of the B.B.C. at which the following memorandum was accepted by both sides:

> The B.B.C. contends that, although the Royal Charter contains comprehensive powers in respect of publishing these powers have not been unfairly used, the criterion being that its publications are pertinent to the service of broadcasting.
> The B.B.C. will recognize and deal with a Committee to be established representing the interests which met the Corporation.
> The B.B.C. is prepared to discuss with the Committee any new publishing proposals, and to consider representations by the Committee concerning existing publications.
> The B.B.C. states that it is not intended that *The Listener* should contain more than 10 per cent original contributed matter not related to broadcasting. The rest of the paper will consist of talks which have been broadcast and comments thereon; articles relating to broadcast programmes and programme personalities, and news of the broadcast service generally.
> The B.B.C. has no intention of publishing any further daily or weekly newspaper, magazine or periodical. It has also no intention to publish books or pamphlets not pertinent to the service of broadcasting.
> The B.B.C., as an evidence of its goodwill, states that it does not intend to accept for *The Listener* more advertisements than are necessary with its other revenue to cover its total cost.[3]

Edward Arnold, Milford and William Longman were appointed to represent the Association on the Committee and at its first meeting in February 1929 they objected to certain of the Corporation's publications, particularly those with numerous illustrations, on the debatable ground that they were likely to have a sale unrelated to broadcasting; and they stated their opinion that 'the whole system of preparing school courses of instruction, having only a nominal connection with Broadcasting Talks, and including the examination and correction of school children's papers on history, geography and other subjects goes far beyond what was

[1] T. R. Colquhoun Dill, 14 November 1928 (*Members' Circular* VII, no. 21, 216–17).
[2] Edward Arnold (President), Humphrey Milford and H. Scheurmier.
[3] *Members' Circular* VIII, no. 1 (January 1929), 3–4.

Broadcasting 87

contemplated in the B.B.C.'s charter'.[1] Whether contemplated or not, broadcasting to schools was to become a new medium of teaching of the first importance, but the printed book was to survive this and other threats to its usefulness.

POST-WAR COSTS

To publishers in general the continuing rise in their manufacturing costs, rather than the expected fall, in the first years of peace must have seemed the greatest menace. Although in 1920 a demand by the Printing and Kindred Trades Federation for a 44-hour week was not pressed, wages rose and printing charges were increased on two occasions in that year. In October Milford was asked by the Federation of Master Printers if he could supply, for consideration by the Joint Industrial Council,[2] evidence to show that the high cost of production was leading to a diminution in the work being put in hand, and the information which he received from forty-five members of the Association showed a large decrease in their output of new books and reprints and a considerable increase in orders placed with foreign printers. The Association also protested to the Master Printers that the second increase (of 5s. a week) in printing and binding wages did not justify an increase of 5% in their charges. In October 1921 and January 1922 there came decreases of 5s. and 2s. 6d. in wages and a reduction in charges averaging 3%; and when in April the Master Printers were discussing with the union the amount of extra work which might result from a further reduction in wages, the new President of the Association, G. S. Williams, could only advise him that, as the increase in printing charges above pre-war figures was nearly 200%, a decrease of even 20% would be unlikely to revive the trade. In March 1920 the Master Binders' Association also gave notice of new charges for the warehousing of publishers' stocks and in the negotiations which followed the representatives of our Association attempted to obtain a scale which would be based on cubic capacity and not on linear measurement, would differentiate between London and country rents, and would continue to give a free initial period. But agreement could not be reached and members had to be left to devise their own solutions by negotiation with their own binders.

There were wide fluctuations in the price of book-paper, a fall from 9d. a pound in January 1921 to $3\frac{3}{4}d$. in May 1922 being quickly followed by a

[1] *ibid.* pp. 12–13.
[2] The establishment of Joint Industrial Councils in industry had been proposed in *The Whitley Report* of March 1917 (the report of a Reconstruction Committee on relations between Employers and Employed, appointed by the War Cabinet under the chairmanship of the Rt. Hon. J. H. Whitley, M.P.) and a J.I.C. for the printing industry had been formed in July 1919.

sharp increase. Two further threats of increased production costs arose also from the Safeguarding of Industries Act of 1921, under which there was a possibility of duties being levied on imported paper and gold leaf. Before the passing of the Act a deputation in which the Association took part was assured by the President of the Board of Trade that the wording of the Act would be so altered that the imposition of a duty on paper could not be brought into effect without publishers having an opportunity of stating their case against it; and the need to do so did not arise. In 1922 a case for the imposition of a duty on the importation of gold leaf from Germany was heard by a committee set up under the Act and was successfully opposed by the Association acting through the London Chamber of Commerce.

STRIKES

The problems of employment and wages in which publishers became involved were not confined to those which were the direct concern of their printers and binders. In February 1924 the Association received a letter from The Mining Association of Great Britain calling attention to the denunciation by the Miners' Federation of the national wages agreement and to the danger, under the Labour Government, of all the manufacturing and distributing industries becoming nationalized, and seeking the co-operation of its sister Association in its fight against an uneconomic minimum wage. It was a prelude to the General Strike of 1926, but before that the publishing trade had a strike of its own. As we have seen,[1] the Association had withdrawn in 1919 from any conduct of labour negotiations and the Book Publishers' Employment Circle, which had been separately constituted in that year, by 1925 had given place to the Book Trade Employers' Federation, including the wholesale distributors as well as publishers who were not necessarily members of the Association. On the other side was the National Union of Printing, Bookbinding and Paper Workers, which had consolidated itself in newspaper and periodical publishing and in the wholesale distribution of books as well as newspapers—at W. H. Smith's and Wyman's—and had thus moved into Simpkin Marshall's, the book wholesalers, and into book-publishers' warehouses. Of the negotiations between these two bodies the Association was officially no more than an interested spectator, but it may not be inappropriate to record its touch-line view of the strikes of 1925–6.

In 1919 with the cost of living index at 105 a minimum wage of 65s. had been agreed for a packer, aged 21, on a 44-hour week. As the cost of

[1] p. 69.

living rose bonuses were added and as it fell they were removed until in 1922 the agreed basis of 1919 was again in operation, though the cost of living index had fallen to 82. The wage was still the same in 1925, with the index at 76. But the union, able to negotiate higher wages in the higher tempo of the newspaper world, claimed that the book trade also could pay more. The Federation quoted the packers' wages agreed by other unions for comparable work—in the cloth and drug trades and in general exporters, ranging from 55*s*. 6*d*. to 64*s*. 8*d*. for a 48-hour week—and stood firm. Although five members of the Federation[1] gave notice that the importance to them of their magazines and periodicals made it impossible for them to act with the rest of the Federation in respect of their book businesses, fifty-nine members pledged themselves to take corporate action to resist the union's demand. When the strike began early in November 1925, there was a cessation of work at Simpkin's and as the employees of the wholesale newsagents were all union members, W. H. Smith's and Wyman's bookstalls and the small newsagent-booksellers were the principal sufferers. Publishers quickly began to get their own books away and before the end of November Simpkin's re-opened.[2] The strike had come too late to disrupt Christmas trade and indeed in its January issue *The Bookseller* reported that business in December had been above average. The Federation, firmly led by its chairman, Bertram Christian,[3] was determined to resist the imposition on book publishing of conditions of employment dictated by the requirements of a different trade. Little progress was made at a meeting on 1 February, and in the third week of that month Simpkin's staff began to return and thereafter was steadily taken back. By March the strike had failed and it was said that outside London it was hardly realized that there had been a strike at all.

It was otherwise when the General Strike began on 4 May. All forms of transport, including the parcel post, were at a standstill and publishers' business was confined to the supply of callers only. With the all-but-total absence of newspapers, booksellers might have expected a good demand for books, but the public seemed to have only two interests, the strike itself and the crossword puzzle, which had made its appearance in the previous year.[4] The General Strike ended on 17 May, but the previous

[1] Cassell's, Eyre and Spottiswoode, Hodder and Stoughton, Hutchinson's, Religious Tract Society.
[2] *Publishers' Circular*, 21 November 1925. The progress of the strike is also reported in the issues of 7 and 14 November and 6 February 1926.
[3] of Christophers and James Nisbet & Co.
[4] 'Crossword puzzles have had a wonderful sale, but had it not been for the sale of dictionaries and companion books we booksellers would have had a thin time. I have heard it from many sources that the sale of general literature—magazines and periodicals particularly—has

strike had left its legacy and in the wholesale houses conditions for restarting work were slow to be agreed.

THE GROUPS

Of the Groups into which the Association had organized itself in 1917 the Educational Group (III) quickly became the most active. Under Humphrey Milford, who became chairman in 1921 upon ceasing to be President of the Association, it began to undertake regular work for the guidance of its members—for instance in the better regulation of the many exhibitions organized by teachers' associations and education committees —and to recommend larger matters of policy (particularly towards the L.E.A.s) to the Council. The conditions to be fulfilled by an Education Authority before it could be recognized as a bookseller, entitled to buy at trade terms, were agreed in 1924;[1] and those Authorities which were so recognized were restrained from supplying municipal and rural libraries to the detriment of local booksellers, and from putting pressure on aided schools to purchase through an Authority's bookstore. The Group also proposed in 1924, in concert with a deputation of the Associated Booksellers, that an agreement limiting the allowable discount on non-net educational books to twopence in the shilling should be drawn up for signature by booksellers and the School Traders' Association, but the Council decided that the underlying difficulties, particularly the unlikelihood of the Associated Booksellers securing unanimity among its members, were too great. In 1927 the Group was formally consulted for the first time by the Board of Education and gave evidence before a committee which was investigating the selection and provision of books for elementary schools. In the same year, after an investigation of the number of school books being published at net prices (which had greatly increased during the war) the Group agreed that all books intended primarily for class use should be published at non-net prices—a practice which has been generally followed since that time.[2] In these formative years the Group owed much to Milford, who was to remain Chairman until 1934, and to Guy Bickers of Bell's and A. J. McDougall of the Scottish firm of that name, who together constituted an advisory committee.[3]

been seriously affected. Fortunately, the craze is on the wane and booksellers should watch their stocks carefully.' Letter to *The Bookseller*, March 1925.

[1] *Members' Circular* v, no. 20 (October 1924), 193.

[2] Except by Macmillan's who after World War II experimented with net-priced school books, but discontinued the experiment after a short time.

[3] In 1924 R. G. Harvey Greenham, a barrister, was appointed as paid secretary of the Group, his fee being provided out of a separate subscription to the Group (£3. 3s. in 1925). The Group also held its meetings in his chambers at 11 King's Bench Walk in the Temple.

The publishers of Juveniles, Reward Books, Annuals, and Toy Books, who constituted Group IV, met irregularly. Trading in a highly competitive range of publications, which attracted also newspaper and periodical publishers who were not members of the Association, they found it particularly desirable to agree regulations which would protect themselves from cutting their own throats and yet difficult to agree not to cut each other's. It was also a branch of the trade which, exceptionally, did much of its business through wholesalers rather than to retail booksellers direct; and each firm had its own list of wholesale firms and wholesale terms. In 1920 the Group agreed an accepted list of wholesalers and attempted to enforce its observance; and again in 1922 it attempted enforcement and tried also to limit internecine competition in the terms allowed both to wholesalers and retailers. In the latter year also mutual mistrust within the Group led to the appointment of a paid Secretary who should not be 'an interested member',[1] and authority was expressly given to him to inspect any members' books. In 1924 a code of rules was agreed, regulating the maximum trade terms allowable and governing conditions of supply including the maintenance of the net prices of Annuals, which were largely dealt in by newsagent-booksellers drawing their supplies from wholesalers. But later in the same year competition from outside the Association necessitated an increase in the wholesale terms on Toy Books published at 1s. or less. It was also the age of 'Bumper' books for children and 1926 brought widespread complaints from booksellers, led by W. H. Smith & Sons, against this monstrosity,[2] printed on a substance approximating to cardboard, and its extravagance in storage space and carriage costs.

Although members of the Fiction Group (II) were invited by the Council in July 1924 to a discussion with representative booksellers from Australia, New Zealand and India of problems of overseas bookselling, and in particular the local retail prices of colonial editions of novels, the Group as such did not meet until January 1926 (its first since 1918) when it agreed to revive itself and elected Capt. Harold Macmillan as Chairman and G. Wren Howard as honorary secretary. At the following meeting in February it reached unanimity about the maximum trade discount to be given on 7s. 6d. novels. Its subsequent activity again concerned overseas problems which will be related in the next chapter.

[1] Harvey Greenham was appointed and the Group subscription was fixed initially at 31s. 6d. This Group also used his chambers for its meetings.
[2] The verbal monstrosity 'to bumperize' was also in use.

THE THREE FOUNDING FATHERS

The years 1926–7 saw the Association, under pressure from a younger generation of leaders, accepting a wider role of responsibility; and to these new aims and opportunities the next chapter will be devoted. It will be appropriate to end the present chapter, which has covered the post-war reconstruction of the Association on foundations laid by its early leaders, with an attempt to give a picture of the personalities of the three founding fathers: Charles James Longman, Sir Frederick Macmillan and John Murray IV. Murray died in 1928 and although Longman lived until 1934 and Macmillan until 1936, their great work for the Association belonged to an earlier era, as indeed it may be thought that in some of their characteristics they did themselves.

To the conduct of his business he devoted constant and careful attention. Six hours a day he regularly spent in his office, except when engaged in meetings or other business elsewhere; and work was frequently prolonged to late hours in his house next door. He wrote an immense number of letters—probably too many—with his own hand; read many MSS himself; and spent infinite pains—often with little recognition—in assisting the authors of works which he published, and in correcting their proofs for the press. Although things sometimes tried his temper, his relations with his employees were of the happiest; and a spirit of harmony and good will always pervaded the 'shop'. He never spared himself, and his untiring energy and unselfishness elevated and vitalized those who worked with or under him...Upright and generous, clear-headed and warm-hearted—a high-minded English gentleman.

So wrote *The Times*[1] of John Murray; and so, without much alteration, it might have written of Longman and Macmillan also.

C. J. Longman formed a trio of close friends with Andrew Lang and Rider Haggard and with the latter he often spent his holidays. It was Rider Haggard too who at the bicentenary dinner of the house of Longman in 1924 paid this tribute to Charles Longman, 'now almost my oldest friend': 'I think of him not as an eminent publisher or as a distinguished prop of commerce, but first and foremost as an English gentleman.' He was keenly interested in natural history and open-air life; and one writer in that field, Richard Jefferies, owed much to his generosity when trouble overtook him and his family. Appropriately, Longman was champion of England at archery.

Sir Frederick Macmillan gave generously of his time to public causes: to the Royal Literary Fund and the Booksellers' Provident Institution;

[1] 1 December 1928.

and he was knighted for his work as Chairman of the Board of Management of the National Hospital for Paralysis. No publisher of his time exercised more personal influence in the book world and as the begetter of the Net Book Agreement he gave this country a charter for good bookshops. 'He was' wrote J. C. Squire,

universally curious and quite free from priggery. He was an ordinary man with ordinary men and a poet with poets. He was a businesslike publisher and rejoiced in being able, because of his commercial success in other directions, to publish occasional magnificent, unprofitable, justifying things. If one had to describe him briefly to someone who did not know him one might say that he was a jolly, kindly, chuckling old gentleman who was the soul of honour and who loved Homer, Shakespeare, Pall Mall, and a good cigar.[1]

In his youth he was a good rider to hounds.

To these three Edwardians—English gentlemen, with the virtues and no doubt some of the faults of that obsolescent character, and tough props of commerce—the Association owes a great debt.

[1] *The Bookseller*, 4 June 1936.

6

1924-1930
New aims and opportunities

THE SOCIETY OF BOOKMEN

The voices of the younger leaders which, as we have seen, had first been heard in 1920, increasingly demanded greater initiative by the Association in seizing the opportunities of a different world and, in particular, a lead to promote the wider distribution of books through greater co-operation between publishers, booksellers, municipal librarians, and authors. In 1924 they began to find expression through the Society of Bookmen, which had been founded in 1921 by the novelist, Hugh Walpole, for the very purpose of bringing together those interested in the writing and distribution of books. At a dinner of the Society in the spring of 1924 it was resolved that co-operative publicity was desirable in the interests of the book trade and the Association was asked to appoint two delegates to an exploratory committee. At its May meeting the Council declined to take part, until some definite scheme was available, but in June it received from two members copies of publicity schemes prepared by advertising agents, of which it had to take notice, and at the July meeting a further invitation from the chairman of the Society, Theodore Byard of Heinemann's, provoked considerable discussion and a decision to rescind the previous refusal. W. M. Meredith and G. S. Williams were then given a watching, rather than an active, brief in the preliminary investigation, but in September, influenced by Stanley Unwin's persistence, they came back with a strong recommendation for the formation of a committee composed of two representatives each from the Associated Booksellers, the Society of Authors, the Publishers' Circle, the Society of Bookmen, and the Association, 'to explore the possibilities of promoting, by collective action, a wider sale of books and an extension of the habit of reading'. To this the Council agreed and again appointed Meredith and Williams as its representatives.

THE NATIONAL BOOK COUNCIL

Early in 1925 the joint committee proposed the formation of an organization to be called The National Book Council, but the Council of the Association found features to dislike in the constitution and recorded that

the Council, while approving generally of the project, are not prepared formally to recommend the N.B.C.'s proposals to their members. They suggest that a fund be raised by voluntary contributions (not necessarily on the scale or in the proportion suggested) to enable a year's experimental work to be carried out by the Executive Committee, who should act privately on behalf of the constituent Associations as a Book Publicity Committee and not as a National Book Council.

It would have been a half-hearted send-off, but in March the Council became assured that the management would be firmly held in the hands of the five sponsoring bodies and that membership would be confined to the members of these five bodies, and executing another *volte-face* it recommended members to give their best consideration to the proposals and agreed to appoint three representatives on the Executive Committee.[1] So was born on 14 May 1925 the National Book Council, to grow in due time—but not without further alternations of hot and cold from the Association, initially representing its principal financial supporters—into the National Book League. With a budgeted income of £750 for its first year, of which two-thirds was to come from publishers and one-third from booksellers, it quickly put in hand the issue of a series of classified bibliographies[2] and the organization of exhibitions and book weeks.

BOOKSELLERS AND THE PUBLIC LIBRARIES

There was a need, in particular, for a more sympathetic understanding between publishers and booksellers, as represented by their trade Associations, of their interdependence and for a joint policy, rather than one dictated by one side only, towards new problems. In November 1919 there had been a joint conference of representatives of the two Associations at which the booksellers had at any rate an opportunity of stating their case for better terms—for general books, $33\frac{1}{3}\%$ on travellers' visits and 25% on other orders; for educational and technical books, 25%; with settlement discounts[3]—and giving their views on other topics such as the

---

[1] G. H. Bickers, Meredith and Williams were the first representatives.

[2] Unwin recalled how the need by Elders and Fyffes for a list of books on the West Indies inspired the idea.

[3] The normal maxima at that time were: $33\frac{1}{3}\%$ without settlement discount on general books; $16\frac{2}{3}\%$ with settlement discount on educational and technical.

recognition of colleges as booksellers and on the timing of the dispatch of review copies. But it had been an isolated meeting and in succeeding years the Associated Booksellers had deplored the lack of consultation; protesting in 1920 against the sale of 'the book of the film' in cinemas; in 1921 against the supply of books on sport to sports shops and in 1924 against the recognition of wireless dealers for the sale of books on allied subjects; and recurrently against the direct supply to municipal authorities of books for their schools and libraries. In 1923 after the recognition of the City of Birmingham's Stores Department in the erroneous belief that its purchases would not include books for its Public Libraries, the Council had given an assurance to the President of the Associated Booksellers that in future recognition would be given only to Local Education Authorities maintaining bookstores.[1] Nevertheless, there remained the reasonable claim of the Public Libraries, as large buyers, to be allowed a discount on their purchases.

In 1925 the question of a discount to Public Libraries was one of the matters referred by the Council to an *ad hoc* Trade Terms Committee, in which G. H. Bickers put forward a novel suggestion for the licensing of book agents, who would include Public Libraries and would be allowed by booksellers a discount of 10%.[2] The Council took up the suggestion and requested the Committee to try to overcome the anticipated objections of the Associated Booksellers. In April the Library Association, whether by accident or with inside knowledge, asked that the rate-aided libraries should be allowed, not to purchase direct from publishers, but to receive a discount, as large purchasers, from booksellers. But when in June the Trade Terms Committee received a deputation from the Associated Booksellers led by their President, G. B. Bowes of Cambridge, it was unable to weaken their conviction that the allowance of any discount would strike at the roots of the Net Book Agreement and their arguments that the work involved in servicing their business with the libraries could not be covered by a reduced margin. Nevertheless the Council felt that the libraries' case could not be summarily dismissed and in July a conference of the three bodies was held. With their President, Scheurmier, in the chair, the Council's representatives listened to the contending arguments—by the booksellers that their service to the librarians was extensive

---

[1] The Associated Booksellers remained unconvinced of the need to recognize L.E.A.s and in November 1927 a deputation was received at a special meeting of the Council of the Association. The booksellers were led by their President, Charles Young, of Lamley & Co. of South Kensington, a bookseller of outstanding character and ability, a friend of Arnold Bennett, and the discoverer of the literary distinction of George Sturt.

[2] *Members' Circular* VI, no. 3 (March 1925), 33.

and costly, by the librarians that they did not need it—and in their concluding remarks they supported the booksellers in deploring any change which might imperil their trade as a profitable livelihood. When later in the month the Council had to make its decision as arbiter, it was that the proposal to allow the libraries a discount under the Net Book Agreement must be rejected. Although Sir Frederic Kenyon accepted the decision on behalf of the Library Association, with the surprising admission that he had never thought that there was a very strong case for a reduction to Public Libraries, the settlement was not to stand for long.

THE CO-OPERATIVE SOCIETIES

In July 1926 it came to the notice of the Council that some Public Library authorities were attempting to get round the Net Book Agreement by becoming shareholders in the Co-operative Students' Bookshops Ltd and so receiving a dividend on books purchased through these bookshops. The Co-operative Societies had been accepted as booksellers for many years and in 1918 the Association's solicitor[1] had fortified it by an opinion that the inclusion of net books in the purchases upon which the dividend was calculated was not an infringement of the N.B.A. But in July 1926 he gave a contrary opinion, whereupon the Manchester headquarters of the Co-operative Union was informed that net books must now be considered as proprietary articles upon which no dividend could be given and was asked for an assurance that this decision would be observed. No reply came and in June 1927 members of the Association were advised to supply the Students' Bookshops with net books at full published prices only. This embargo brought a deputation from the Co-operative Societies to a meeting with the General Purposes Committee in October, at which Sir William Beveridge,[2] who led the deputation, gave an undertaking to recommend to members of the Students' Bookshops at a General Meeting an alteration of the Rules of the Society which would exclude profits on net books from the surpluses available for distribution, and the deputation undertook to stand or fall by the recommendation. As an act of good faith the General Purposes Committee agreed to an immediate lifting of the embargo and by February 1928 the formal alteration of the Bookshops' Rules had been made.

[1] C. R. Rivington, a member of the publishing family of that name.
[2] At that time Director of the London School of Economics; later Master of University College, Oxford; the creator of the comprehensive scheme of social security enacted in 1946-8.

1924–1930 : *new aims and opportunities*

THE VISIT TO GERMANY AND HOLLAND

In 1926 there came new impetus towards organized co-operation between publishers and booksellers, once more from the Society of Bookmen. In April the Association received a letter from Maurice Marston,[1] the secretary of the Society, reporting that a sub-committee had been studying the applicability to English conditions of the system of a clearing house for the dispatch of orders which was in use in Germany, Holland and Sweden and proposing that its operation should be studied on the spot by representatives of the Society, the Publishers Association and the Associated Booksellers. The Council nominated C. S. Evans[2] and when, after the return of the working party from Germany and Holland, the Society requested the appointment of additional representatives to survey the report, it selected G. S. Williams, G. H. Bickers, and Harold Raymond.[3] When the Committee of Survey met in November, with Stanley Unwin (who had led the working party) in the chair, the representatives of the Association said that they had been appointed to consider what could be achieved by better co-operation in the book trade, rather than to co-operate in the drafting of a constitution for the proposed clearing house. But when the other members of the committee and especially those representing the Associated Booksellers urged the importance of forming a 'British Book Trade Council' for its general usefulness and in particular as vital to the operation of a clearing house, the Association's representatives argued that its work could equally well be done by liaison officers between the two Associations or by a joint committee. The Council upheld the view of its representatives and when the Committee of Survey met again on 13 January 1927 it agreed, with some reluctance but *nem. con.*, that the possibilities of greater co-operation between the two branches of the trade, including the formation of a clearing house, should be examined by a joint committee. A week later the Council maintained its position and resolved unanimously

that the Associated Booksellers be invited to set up with this Association a joint committee, consisting of twelve members of each body, to examine and report in confidence to their respective executives on the possibilities of greater co-operation between the two Associations, for the purpose of improving the organization of the trade, and that the procedure shall be settled by the officers of the respective Associations before the committee is called into being.

[1] Marston was also secretary of the National Book Council.
[2] Heinemann.
[3] Chatto & Windus.

So came into being 'The Joint Committee'—as it was called—and although the proposed clearing house had been submerged, and was indeed to be allowed to sink, a larger voyage of discovery had begun.

'THE JOINT COMMITTEE'

Rules for the procedure of the Committee in relation to its parent bodies and provision for its secretarial and other expenses were duly agreed by the officers, the members were appointed[1] and when the Committee met for the first time on 10 March 1927 it elected Scheurmier, the President of the Publishers Association, as its chairman and G. B. Bowes, the President of the Associated Booksellers, as vice-chairman. Bowes became chairman after Scheurmier found it necessary to resign in November and it was under him that the effective work of the Committee was done. That work comprised an Interim Report in March 1928, embodying the findings of three sub-committees; a second report in December 1928, making recommendations upon some of the findings; and a further report with the remaining recommendations in May 1929. In all it provided a lasting charter for co-operation between the two branches of the trade and for some publishing elders its acceptance required a new understanding of the rights of booksellers.

Working through the three sub-committees, the Joint Committee made a survey of the trade under three headings: getting orders; filling orders (distribution); trade organization and practices. From the premises that in comparison with 1890 the year 1926 had seen an increase of about 150% in the number of publishers issuing books and an increase of about 123% in the number of books published and that the growth of population and compulsory education in the United Kingdom might be assumed to have increased the reading public by about the latter percentage, it began by analysing the objects of publicity, the effectiveness of the methods used, and the respective responsibilities of publisher and bookseller. It noted the lack of uniformity in the arrangement of publishers' catalogues and the lack of co-operation by booksellers, some of whom had no classified lists of

[1] The members were: *for the Publishers Association:* G. H. Bickers (G. Bell & Sons); Jonathan Cape; C. W. Chamberlain (Methuen & Co.); G. Duckworth; C. S. Evans (Heinemann); W. Longman; H. Raymond (Chatto & Windus); G. C. Rivington; W. Symons (Blackie & Son); Stanley Unwin (Allen & Unwin); G. S. Williams (Martin Hopkinson), Vice-chairman; G. Wilson (A. & C. Black); *for the Associated Booksellers:* H. E. Alden (Simpkin Marshall); G. B. Bowes (Cambridge), Chairman; F. Brown (Educational Supply Association); F. A. Denny (London); G. Foyle (London); F. J. Hanks (Blackwell); D. Roy (W. H. Smith & Sons); A. Stevens (Times Book Club); E. Story (York), later J. Norman Read (Bolton); J. Ainslie Thin (Edinburgh); T. C. Ward (Midland Educational Co.), later W. S. Sisson (Nottingham); C. Young (London). *Secretary* Maurice Marston.

their customers, in the distribution of prospectuses, with the consequent encouragement to publishers to find other channels of distribution. It put the jacket, which had only recently emerged from the chrysalis of the dust-cover, into three categories—illustrative, decorative, informative—and commended the use on it of the descriptive 'blurb' rather than the self-laudatory 'puff'. From replies to a questionnaire sent to booksellers maintaining circulating libraries it concluded that only 23% of book-borrowers were book-buyers, that 82% of the books borrowed were novels, and that the habit of book-borrowing was not generally conducive to buying. Finally in its survey of the getting of orders the Committee suggested improvements in publishers' service to booksellers through their travellers, deplored the wastage of effort in the advertising of new publications to the trade owing to the existence of more than one trade periodical, and reported that it was unable to recommend the establishment of a clearing house for orders in the face of evidence critical of such an organization. It was to be many years before it came to be seen, as it now is, as an indispensable part of the organization of the trade.

In its survey of the channels and methods of distribution the Committee did its most important work. As regards the granting of trade terms it could see no clear principles in the Publishers Association's list of those 'recognized', which included Government Departments, Local Education Authorities, Co-operative Societies, certain schools and colleges, religious, political and philanthropic societies, professional associations, certain libraries, and even some individuals who could not be classified. 'There is no difficulty' the Committee said, 'with general booksellers who have a shop open to the public all the year round, but uncertainties begin to arise in the case of: (i) Specialist shops. (ii) Those having no shop. (iii) Those whose shops are not open to the public all the year round. (iv) Those doing direct mail order business. (v) The general village shop. (vi) Societies, Institutions, etc.' The Committee was of the opinion that there should be two classes of bookseller: 'A' booksellers, entitled to trade terms on all publications; 'B' booksellers, entitled to trade terms only on particular publications pertaining to their trade, profession or organization, with the rider that it would be in the interest of the trade generally that as much business as possible should go through 'A' booksellers. The Committee proceeded to consider methods and costs of delivery, including an analysis of the standard contract of the carriers, Messrs Sutton, who were predominantly used by booksellers, and to make suggestions for regular publishing days for new books and for the dispatch of booksellers' pre-publication orders in time to arrive on the preceding evening. Of the essentiality of the Net Book Agreement to the health of the whole trade the Committee had no doubts and finding evidence of some opportunities for its evasion, it recommended that the agreement should be accompanied by a code of interpretations and that it should be extended to cover sales made in Great Britain by exporting booksellers and by commission agents buying for institutions abroad. To this important recommendation and to the outcome of the Committee's proposals for the systematization of the granting of trade terms at home we shall return.

In its survey of trade organization the Committee noted that the trade had no less than five periodicals catering for its needs—*The Publishers' Circular*, *The Bookseller*, *The Newsagent*, *The National Newsagent* and *The Clique*—two competitive annual publications listing the year's books—*The English Catalogue* and *Whitaker's Cumulative Book List*—and no adequate directory. Five periodicals inevitably involved much waste of reading time and twelve features which the Committee considered essential were not adequately covered in any one of them; moreover, all except *The Clique* could be purchased by any member of the public, so that discussion in them of purely trade questions was sometimes inadvisable. It was thought that an official trade paper was desirable and that it would be financially profitable, for from Stanley Unwin's first-hand experience of the German book trade the Committee knew that the Börsenverein derived its strength from the advertisement revenue of the *Börsenblatt*, its official journal. Upon this delicate question of an official trade paper the Committee presented a separate, confidential report to its parent Associations, the unsuccessful outcome of which I shall relate. The provision of a Book Trade Directory was among the main recommendations of the Committee to which effect was given.

In its investigation of trade practices the Committee endorsed the longstanding recommendations of the Publishers Association upon bibliographical descriptions: that the title-leaf of every book should bear the date of the year in which the impression or the re-issue, of which it forms a part, was first put on the market; that when stock is re-issued in a new form, it should be described as a re-issue; and that the date at which a book was last revised should be indicated on the title-leaf. On the controversial question whether bindings which are too narrow to be lettered across the spine should be lettered up or down the Committee recommended that the standard practice should be that when the volume stands (the right way up) on the shelf the lettering reads from bottom to top. But lettering from top to bottom continued to have its adherents and was indeed preferred by the 1948 Book Trade Committee.

The Committee also recommended a standard practice, which is still in common use by publishers, for the giving of notice to booksellers of the impending publication of revised and cheap editions and of remainders and for the calling in and replacing, or giving credit for, booksellers' current stocks.[1] The Committee had also something to say about partial remaindering: 'the practice (of a few publishers) of selling small portions of the remaining stock of a book at remainder prices to a few privileged firms, and continuing on ordinary terms to others, was found to be the cause of much ill-feeling among booksellers as being obviously unjust. The practice, fortunately, does not seem to be at all widespread'— a practice, nevertheless, which might then have been more severely condemned, as subsequently it was.

THE STANDING JOINT ADVISORY COMMITTEE

The recommendations of the Joint Committee for the systematic consideration of applications from booksellers and other traders in books in the British Isles for entitlement to buy at trade terms, whether for sale at home or for export, was accompanied by a recommendation that all applications should be referred to a standing Joint Advisory Committee, consisting of four representatives elected by the Publishers Association and four by the Associated Booksellers, which would also undertake a review of those already recognized. 'Recognition' would be given as general booksellers, and other traders wishing to carry books relative to their own trade would also be recognized for supply at trade terms, although not necessarily full terms, in respect of such books only, as would department stores wishing to carry books for seasonal trade only. Export traders in the United Kingdom and the Irish Free State would be classified either as export booksellers (buying for overseas booksellers), who would be supplied on trade terms, or export commission agents (buying for firms and private individuals resident abroad who were not purchasing for re-sale), who would be given a commission not exceeding 10% of the published price; and both would be required to sign the Net Book Agreement. These recommendations were approved by both associations with some amendments, in particular to meet the insistence of the Council of the Publishers Association upon three points: that the establishment of the Joint Advisory Committee should in no way challenge or alter 'the publisher's admitted right to find his own channels of distribution'; that applications from Local

---

[1] These recommendations together with all the other recommendations approved by the Publishers Association and the Associated Booksellers are to be found in *The Publisher and Bookseller* of 5 July 1929.

Education Authorities should not be referred to the J.A.C.; and that the J.A.C. should report to the Publishers Association only and not to the Associated Booksellers. Even though the former was to have the sole and final say upon the recommendations of what was to be no more than an advisory committee, Sir Stanley Unwin recalls the difficulty which he, as the principal spokesman for the Joint Committee, had in winning over Sir Frederick Macmillan, the most powerful objector.[1] William Longman also made a notable contribution to the work of the Joint Committee, which was recognized by his election as President of the Association in March 1929,[2] and when in June the Council appointed its first representatives on the J.A.C., it selected Longman, Unwin, G. H. Bickers and Harold Raymond. The first meeting of the J.A.C. was held later in that month and in the following September the Council approved its first list of recommendations for the granting or refusal of trade terms to general booksellers, 'other traders', and export booksellers. The J.A.C. continues, virtually unchanged in status and functions to this day, although publishers still retain their individual right to give or withhold trade terms as they choose, regardless of any listings by the J.A.C.

INTERPRETATION OF THE NET BOOK AGREEMENT

To stop evasions of the Net Book Agreement the Joint Committee recommended, and the two Associations agreed, not only that it should apply to all export sales executed in the British Isles, but that it should be accompanied by 'decisions and interpretations'. Accordingly in July 1929 the Publishers Association ruled that:

it will be deemed to be a breach of the Net Book Agreement if any bookseller, store or circulating library

(i) offers or gives any consideration in cash to any purchaser, except under licence from the Publishers Association.

(ii) Offers or gives any consideration in kind, e.g. card-indexing, stamping, reinforced bindings, etc. at less than the actual cost thereof to the bookseller.

(iii) Offers by advertisement or otherwise to pay or partly to pay postage or carriage on an order for books to the value of less than £2.

(iv) Offers or supplies non-net books at less than the actual cost thereof to the bookseller himself, unless and until some agreement on further limitation of discount on non-net books has been reached.

(v) Offers to pay or pays railway fares.

[1] *The Truth about a Publisher*, p. 376.
[2] He succeeded Edward Arnold who had been invited by the Council in September 1928 to be President until the next Annual General Meeting, W. M. Meredith having resigned for reasons of health.

(vi) Offers second-hand copies of net books to the public at reduced rates, whether by postcard, leaflet, circular, or any other form of publicity within six months of publication, even though the book itself is not to be reduced in price till the six months have expired.

The Restrictive Trade Practices Act of 1956 required a complete revision of the Net Book Agreement and the abandonment of some of the 'Decisions and Interpretations', e.g. those numbered (iii) and (v), but its nature is substantially unchanged.

THE LIBRARY AGREEMENT

The reasonable claim of the Public Libraries to be permitted to buy their books at less than the published prices had been rejected by the Association in its support of the booksellers in July 1925, but it could not remain unrecognized; nor were local booksellers generally retaining the business, which was increasingly being secured by library contractors. The Rules of the Net Book Agreement were being defied and when the Report of what was known as the Kenyon Committee on *Public Libraries in England and Wales* was presented to Parliament in May 1929 it included the following exhortation:

it appears to be notorious that evasions of the existing rule are in fact practised to a considerable extent. Booksellers themselves offer arrangements to public libraries, which amount to evasions of the net book agreement, and public libraries in some cases avail themselves of them. This practice raises problems of a moral character which had much better be avoided. It is infinitely better that there should be some accepted and legitimate agreement than that any opening should be left for imputations on the good faith of either librarians or booksellers.[1]

It was the irregular allowance of discounts to Public Libraries which was striking at the roots of the Net Book Agreement, rather than their regularization which Bowes had seen as the menace two years before. There were two alternatives: that, as in the United States and Canada, the libraries should be supplied direct by publishers at a discount or that the booksellers should be permitted so to supply them. The former would have been the more profitable to publishers, and would have had the additional advantage of bringing them into direct contact with the librarians, but by the publisher members of the Joint Committee at any rate it was believed that, for the good of the trade as a whole, library business should be retained, and if possible regained, by local booksellers.[2]

[1] Cmd 2868, H.M.S.O. (1927), para. 612.
[2] The alternatives and subsequent misunderstanding of them were recorded by Stanley Unwin in a letter to *The Publishers' Circular* of 7 August 1937.

When the Joint Committee accordingly recommended in its interim report that public libraries should be licensed, as book agents attached to named booksellers, to receive a discount, the Council acted without waiting for the full recommendations of the Committee and in May 1929 appointed William Longman and Stanley Unwin to negotiate with the Library Association. In July they were able to report that the basis of an agreement between the Publishers and Library Associations, acceptable also to the Associated Booksellers, had been reached; in October the formal agreement was submitted to the full membership of the Publishers Association at a special General Meeting; and on 12 November 1929 it was signed and came into force. It provided that libraries in Great Britain and Ireland giving free public access and spending not less than £100 a year on new British books should be licensed to receive a commission of 5% from booksellers named in the licence and those spending not less than £500 a year a commission of 10%.[1] Before the agreement had been in operation for a year the Council of the Library Association began to press for its amendment to give all licensed libraries the 10% commission irrespective of their annual expenditure or at least to extend it to the smaller libraries spending not less than £100 a year. The latter amendment was conceded by the Associated Booksellers at their annual conference in June 1932 and subsequently approved at a General Meeting of the Publishers Association in the following November. On these terms the agreement continued throughout the period of this narrative.[2]

THE OFFICIAL TRADE PAPER AND TRADE DIRECTORY

Two other needs to which, as we have seen, the Joint Committee drew attention were a single, official trade periodical and a trade directory. While the Joint Committee was still at work in October 1927 the Council set up a Trade Paper Committee to consider what steps should be taken to provide the trade, both wholesale and retail, with a practical and efficient periodical dealing with all matters affecting it. They appointed to it the three Officers (Meredith, Scheurmier and G. C. Rivington), T. Byard, Stanley Unwin, and G. S. Williams, who was made chairman;

---

[1] It also provided initially that no commission or discount would be allowed on books on which the bookseller did not receive a trade discount of more than twopence in the shilling plus 5%. Although in 1931 the Associated Booksellers offered to waive this condition, the scheme does not now require booksellers to allow any discount at all on books on which they themselves receive a trade discount of less than twopence in the shilling plus 5%.

[2] The agreement ended in 1964 when the constitution of the Library Association had ceased to include corporate (as distinct from professional) members. The substance of the agreement is still preserved in the Library Licensing Scheme operated by the Publishers Association on the advice of a J.A.C. which no longer includes any representative of the Library Association.

and the Committee co-opted three booksellers, Charles Young, the President of the Associated Booksellers, G. B. Bowes, and F. J. Hanks, who were members of the Joint Committee. In September 1929 the committee (which had been reconstituted to control the official trade periodical) was asked also to consider the publication of a trade directory.

The Trade Paper Committee met six times and submitted its recommendations in March 1928. It began by considering whether a trade paper should be privately controlled or controlled by one of the two trade associations or controlled by both; and preferring the authoritative to the controversial it unanimously favoured the last. It then considered whether it would be better to launch a new paper belonging to the two associations or to try to obtain control of one of the existing ones; and, again unanimously, it chose the latter. Of the five papers noted by the Joint Committee, only two were primarily directed towards book-publishers and booksellers: *The Publishers' Circular and Booksellers' Record* and *The Bookseller*. *The Publishers' Circular* had been established by Sampson Low at the instigation of a Committee of leading London publishers in the year in which Queen Victoria came to the throne and he had subsequently been joined in its control and in the partnership of the firm of Sampson Low, Marston & Co. by Edward Marston, with whom H. M. Stanley and R. D. Blackmore found friendship in the publication of *Darkest Africa* and *Lorna Doone*.[1] In 1927 *The Publishers' Circular* was managed and edited by E. W. Marston, grandson of Edward and son of R. B. Marston, who had held the office of Honorary Secretary of the Publishers Association from 1896 to 1904. *The Bookseller* had been started in 1858 by Joseph Whitaker of *Almanack* fame and G. H. Whitaker, the third editor of that name, was in charge in 1927.

The Committee decided to approach first *The Publishers' Circular* as the older of the two papers and after several meetings with E. W. Marston it put to him for the consideration of his board a scheme for the acquisition of the control of the paper by the formation of a new company in which the ordinary, controlling shares would be held by the two associations, whose joint investment was estimated at not more than £400. The board found itself unable to recommend the scheme to the shareholders and made a counter-proposal under which the two associations would adopt the paper as their official organ and appoint the editor, but the Associated Booksellers only would appoint a director to the board. This the Committee in its turn rejected: first, because the editor would be ultimately under the control of a board representing shareholders and not the associations;

[1] see F. A. Mumby, *Publishing and Bookselling* (Revised ed. 1949), pp. 276-7.

secondly, because the only means by which the associations would share in the financial benefits of their support was through the purchase of such shares in the company as might come into the market. The Committee had it very much in mind that book-trade papers in other countries controlled by trade associations were a source of income to be used in promoting the interests of the trade, and it accordingly put forward a revised scheme for acquisition of the control of the paper. But Marston's board would consider no alternative to its own scheme and fixed a time-limit for its acceptance; and before the time expired the Committee received an approach from Cuthbert Whitaker on behalf of the proprietors of *The Bookseller*. With him the Committee agreed: that the two associations should adopt *The Bookseller* as their official organ and that its title should be changed to *The Publisher and Bookseller*; that the two associations should appoint the editor and be free to change the arrangement of the paper and its format; and that the proprietors should, subject to certain reasonable limitations, pay all expenses and hand over to the two associations any profits after the deduction of a commission of 20% of the revenue. An agreement on these terms was recommended by the Council and approved at the Annual General Meeting in March 1928. A new Trade Paper Committee was appointed to control the editorial policy and methods, upon which the Council nominated as the Association's representatives Stanley Unwin (who became chairman), Byard, and Meredith until his resignation of the Presidency when he was succeeded by Harold Raymond. G. S. Williams, who had been the chairman of the original Trade Paper Committee, was appointed editor and the first number of *The Publisher and Bookseller* appeared in April 1928. The trade had got its official runner, but in spite of repeated requests to support it and of Whitaker's generosity in foregoing their commission in the first year, publishers increasingly put their advertising stakes on the other horse. A perverse spirit of individualism seemed to provoke them to revolt against official direction, but it must also be said that the ideal editor had not been found in Williams, who must be judged from his record as a leader of the Association to have been conservative by temperament and, at the age of fifty-eight, insufficiently forward-looking for this new assignment. It must also be doubted whether what a trade paper gained in authority and accuracy from official backing would not under any editor be more than counter-balanced by its inability to criticize and to take sides in matters of controversy. To these reasons was added the great trade depression of 1931-2 and in 1933 Whitaker's exercised their right to terminate the agreement. In October *The Bookseller* re-appeared under its old name, but under a new editor, Edmond Segrave, then aged twenty-eight. The paper had lost the weight of official backing, but Whitaker's bold appointment was to give it an editor with a strong sense of responsible impartiality to both sides of the trade and with a polished, deadly rapier in his hand.

In the meantime, in their continuing belief in the need for an official paper, the two Associations were proceeding to run the other horse. A financially attractive offer from *The Publishers' Circular*, under which the Associations would have an equal interest with the shareholders in the profits of the paper, was accepted in 1933 and a new Trade Paper Committee, with an indefatigable chairman in G. Wren Howard, began in co-operation with the editor, E. W. Marston, to try to introduce improvements in content and in format. At first the Associations benefited financially, but once more publishers began to back the unofficial runner with their advertising money; and in 1936 an *ad hoc* committee of publishers and booksellers was set up to consider how the paper could be more drastically improved. Once again the editor was not cast for the part; liveliness and official impartiality do not easily go hand in hand, but he was accused of a lack of both. For the year 1937–8 no profit was received by the two Associations, but nevertheless the report of the joint Committee, based on criticisms elicited by a questionnaire, proved to be unacceptable to the proprietors and towards the end of 1938 it was mutually decided that the operation of the agreement should be suspended for one year and that unless a new understanding could be reached during that year, the agreement should automatically determine. It lapsed at the end of 1939.

It was with Whitaker's also that the associations entered into an agreement for the publication of a Trade Directory. The Trade Paper Committee estimated that the cost of production could be assured from the revenue from sales and advertisements, but that the cost of compilation would have to be underwritten by the raising of a sum of £500. Of this the Associated Booksellers offered to contribute half and the publishers' half was met by loans of £10 from twenty-five firms, the whole to be repayable without interest as a first charge on the receipts after the expenses of production had been recovered. An agreement between the two Associations as proprietors and Whitaker's as publishers was signed in September 1930; and the first official Trade Directory was published in 1933.

Co-operation had been the key-note of the 1920s and the achievement of it was celebrated at a dinner given by the Association on 19 November 1929 at Stationers' Hall. Of the guests representing allied associations the President of the Associated Booksellers had pride of place, and the men of science and letters present included A. S. Eddington, J. H. Jeans, Rudyard Kipling, Dean Inge and G. S. Gordon. With William Longman in the chair, there was a series of long speeches, lightened by sketches by comedians. Of this double entertainment an act by Arthur Askey alone has survived in one memory.

COMMONWEALTH BOOKSELLERS

As co-operation between publishers and booksellers increased at home, so requests for a better understanding of their problems came from booksellers in the Commonwealth, and particularly from Australia and New Zealand. Then as now these two countries were blessed with bookshops of exceptional quality. Their quality came in part from the comprehensive stocks which they carried because remoteness from their sources of supply made quick replenishment impossible; heavy stocks increased their risk; greater risk and heavy transport and landing charges required them to increase their retail prices above the U.K. published prices; and the scale of 'marking up' brought prosperity and an organization determined to preserve it. In 1922 T. Maskew Miller, on behalf of the booksellers of Cape Town, requested the Publishers Association to fix a South African retail price for colonial editions of novels and to protect South African booksellers from underselling by London booksellers, but the Council was unsympathetic and the South African trade was not sufficiently organized to attempt to introduce a Net Book Agreement of its own. That, in effect, was what Australia was to achieve after some six years of persistent pressure.

AUSTRALIAN TERMS AND PRICES

In 1923 the New Zealand Booksellers' Association made the same complaint as Maskew Miller had done in the previous year and got the same reply. But a conference of Australasian booksellers meeting in Sydney in May 1924 dispatched a deputation to London to voice their grievances. The deputation, which was received by the Trade Committee and had a meeting also with the Fiction Group, was joined by a representative of Maskew Miller, but as the South African voice was not backed by an organized association it could not be effective in this or subsequent negotiations. The Australasian demands were that booksellers in the two countries should be supplied at half price, so that they could sell at the U.K. published prices; that penalties for underselling should be enforced from London; that the retailing of colonial editions should be confined to overseas booksellers and be denied to exporting booksellers in the U.K.;

and that schools, agents buying for schools, and Government Departments should not be supplied direct by publishers. As the discussions advanced it seemed to be the sale of colonial editions of novels with which the deputation was most concerned and the Council agreed to recommend members not to supply them to London exporters except on the understanding that they would not be retailed at less than the prevailing prices recognized by colonial booksellers' associations. But in Australia booksellers remained unsatisfied by this limited concession and in their demand for the enforcement of higher Australian prices for all imported books they had the support of Australian publishers who saw in it an easing of their own competitive position. In January 1925 news reached London that the booksellers of New South Wales had decided to refuse to do business with any publishers' representative who called also on department stores unless the stores were compelled by British publishers to maintain the advanced prices agreed by the booksellers. After a fruitless meeting in Sydney between representatives of the booksellers, the stores, and publishers' local representatives, C. F. Clay, then President of the Association, sent in March a strong statement of the reasons why the enforcement in London of a Net Book Agreement to maintain Australian prices fixed by booksellers would be unacceptable in principle and impracticable in operation. Clay's letter was read to a conference of the Associated Booksellers of Australia and New Zealand in May and in his reply the President, George Robertson, said that 'nothing more callous was ever penned' and that if it represented the views of a majority of the Association, good bookselling in Australia was doomed.[1] The conference reiterated the full demands which had been made by the deputation, including in particular an Australasian net price schedule of which a copy was enclosed, together with an offer to contribute up to £1,000 a year towards the cost of its enforcement.

It is unnecessary to recount the lengthy correspondence and all the comings and goings in person during 1925–8, in which the Association in London insisted that the scheduled prices were higher than booksellers in some parts of Australia found it necessary to charge—and higher than those charged by Australian publisher-booksellers for their own competing publications—and that the Australian Booksellers' Association must first become fully representative, bringing in the stores and ending the embargo upon publishers' salesmen who called on them; to all of which the booksellers in Australia, and to a lesser extent in New Zealand, maintained their opposition. In October 1927 the Council set up an Australasian

[1] The letters are printed in *Members' Circular* VI, 39–41 and 80–4.

Committee[1] to renew negotiations and in March 1928 news came from the Association of British Publishers' Representatives in Australia that it had been asked to co-operate in forwarding some movement towards the settlement of the price question. The prospect of a fresh deputation from Australia and New Zealand was welcomed by the Committee and in July 1928 there arrived a party of seven, representative of booksellers, stores, and publishers' representatives. After three meetings, with intervening reference to the Educational and Juvenile Groups and to the full membership of the Association at a special General Meeting, agreement was reached upon a statement of terms for the sale of books in Australia, with a schedule of net prices, and for its policing by the Association of British Publishers' Representatives. It was hoped that the necessary signatures of Australian booksellers would be obtained by Christmas, so that the agreement might come into force there and, if the New Zealand booksellers decided at their annual convention to adopt the same scheme, in that country also in January 1929. In the event New Zealand could not agree to adopt the scheme and it went forward for Australia only. Signature by Australian booksellers and subsequent signature by members of the Association took longer than had been expected, but 'The Statement of Terms for the sale of Net Books in Australia' came into force on 1 January 1930.

AUSTRALIAN ECONOMIC DIFFICULTIES

Scarcely had the 'Statement of Terms' come into operation when the fluctuation of the Australian pound, following the disastrous fall in wool prices in 1929, put the schedule of prices into jeopardy. In the hope that the fluctuation was temporary, it was agreed at a special General Meeting of the Association in August 1930 to support the schedule as minimum prices only and, with that concession, to insist that no help could be given from the U.K. to mitigate the effect of the adverse rate of exchange. When, in July 1931, the Victorian branch of the Australian Booksellers' Association suggested a new schedule based, not on the U.K. published price, but on the trade price to the Australian bookseller and varying according to the kind of book and the terms of the particular publisher, it was reminded that any increase on the agreed schedule must be uniform and be confined to the exchange and other landing costs. At the end of 1931 the rate of exchange was pegged at £125 Australian to £100 sterling,

[1] The Officers (Meredith, Scheurmier, Rivington), Sir F. Macmillan, W. B. Cannon (Oxford University Press), C. W. Chamberlain (Methuen), W. Symons. G. S. Williams and Stanley Unwin were added to the Committee later.

but agreement on a new schedule was not in sight; and the Australian Committee was merged into the Export Committee until a need for its re-emergence arose. Schedules submitted in 1933 and 1937, which were open to the same objection as in 1931, were not thought to offer a basis for negotiations.

The devaluation of the Australian pound was not the only result of the impact upon the Dominion of the world-wide trade depression which began in 1929. Not only was the duty on advertising material printed overseas, which had first been introduced in 1908, increased by 10% and a landing duty of $2\frac{1}{2}$% imposed on overseas publications, but the Australian printing industry requested the Minister of Customs to impose a higher protective tariff on all overseas printed matter, which it contended would give employment for 10,000 Australians, and to amend the Australian copyright law by the enactment of a 'manufacturing clause' similar to that in force in the United States. Against these greater measures of protection the Associated Booksellers of Australia protested vigorously and successfully; in a report published in September 1931 the Tariff Board concluded that a duty on books or periodicals would cause unemployment in the bookselling trade exceeding any additional employment in the printing industry, that it would be a tax on knowledge, and that it would give rise to more resentment in the United Kingdom than any other tariff which the Commonwealth could devise.[1]

[1] In 1927 the Canadian Lithographers' Association had made a similar request for a high protective tariff: on novels and on unbound books of all kinds 25% (15% preferential rate on importations from Great Britain); non-fiction, if bound, 15% (10% preferential). The demand was rejected by the Tariff Board in Ottawa on evidence organized by the Publishers Association in conjunction with the British Federation of Master Printers.

7
1931-1939 (1)
The Great Depression, markets and rights

The 1930s began in a world-wide economic blizzard and ended in world-wide war. By 1928 the unsound structure of the European economy, propped up largely by loans financed by American investors, had begun to crumble and American investment at home had produced a great speculative boom in stocks and bonds. In 1929 confidence in endless prosperity and endlessly rising stock prices had given way until in October there had come the great Wall Street crash, when in one week 240 securities had declined in market value by $16,000 million. 'The Great Depression' had begun. Britain's total exports fell from £839 million in 1929 to £658 million in 1930 and £455 million in 1931; and the number of unemployed in 1930 rose from 1½ million in January to 2¼ million in December. The withdrawal of gold became an avalanche in 1931 and in September Britain abandoned the gold standard and the pound fell from $4·86 and after fluctuations settled for a time at a rate around $3·50. An Import Duties Act followed in 1932.

It was thus a decade of stagnation in international trade from which books did not escape, and in which publishers found themselves contending with obstacles to foreign trade then new to them, although regularly familiar to their successors. At home, perhaps because financial depression curtailed more costly forms of expenditure, the book trade held its own better than it expected. 'Economic recovery' wrote Stanley Unwin in March 1935, the second year of his Presidency,

has, we are told, to some extent been achieved in this country, but whether this be so or not the book trade would be ungrateful to complain of its position. The trade has not suffered to anything like the same extent as have many others, and during the year under review a definitely optimistic note was sounded, particularly towards the close of 1934. Such progress as has been made, however, has been confined to the home trade, and unfortunately has to some extent been offset by the fact that export business has become increasingly difficult. Exchange restrictions, tariffs, certificates of origin, piracies and so on involve an ever increasing expenditure of time.[1]

In keeping their export business alive in spite of these barriers publishers were to get much help later in the 1930s from the British Council, which under Unwin's persuasion was quick to see that books could be a powerful instrument for the projection of Britain's image abroad. That trade at home progressed in a stagnant economy was due in part to new ideas, some of permanent, some of temporary significance: the launching of book clubs, the invention of Book Tokens, the mushroom growth of the so-called 'Twopenny libraries', and the promotion of the *Sunday Times* Book Exhibitions and other book weeks under the management of the National Book Council.

DEVALUATION AND IMPORT DUTIES

The fall in the pound sterling in September 1931 led to an immediate increase in the cost of binding, for which not only the cloth but most of the other raw materials—dyes, starch, oxide and gold leaf—were imported; and a further increase followed the Import Duties Act of 1932. In September of that year the principal firm of cloth manufacturers claimed that its costs had consequently risen by 25% in the preceding twelve months and although the Association had to accept that claim, it was forcibly reminded that that one firm had a near monopoly of the supply and that competition from newcomers should be encouraged.[2]

That foreign books, bound and unbound, were exempted from the Import Duties Act during its passage through Parliament was largely due to representations made by the Association to the Chancellor of the Exchequer and the President of the Board of Trade. But electrotype and stereotype plates, and blocks for the printing of illustrations, were covered by the Act and the dutiable figure included not only the cost of the plates and blocks themselves, but also fees or royalties payable for their use. By July 1934 the Association was able to persuade the Customs authorities that royalties which might subsequently be due should not be assessed and that the duty should be calculated only on the invoiced price of plates and blocks.[3] The Act also subjected to duty imported catalogues without exception and it was December 1934 before the Association could secure the separation of learned catalogues from trade catalogues and their addition to the free list.

---

[1] *Report of the Council, 1934–1935.* [2] see pp 156 ff.
[3] This point was again a matter of argument with the Customs in the 1950s.

The circulation of British books abroad was limited by the various restrictions which many countries found it necessary to introduce to defend their economies against the great depression. Except in the United States and Canada books continued to be admitted free of Customs duty almost throughout the world.[1] It was therefore only in Canada that British books benefited from the Ottawa Agreements of 1932 under which the countries of the British Empire gave imperial preference, generally by discriminatory tariffs against non-Empire goods; and in Canada the reduction in the import duty by 5% was more than offset by the imposition of an excise tax of 3% and an increase of the sales tax by 3%, at the absurdity of which the then President of the Association, Bertram Christian, and G. Wren Howard protested in person during a visit by the Canadian Prime Minister in June 1933. In Australia the threat of an import duty in 1933 was opposed by the Association and by publishers' representatives in the Dominion and was defeated by local agitation by the cultural and trade interests affected; and by April in the same year books were free also of the internal taxes which had been imposed after the devaluation of the Australian pound in 1930. Similarly in New Zealand a sales tax on books proposed in January 1933 was withdrawn.

Of European countries Iceland alone proposed in 1935 to include books in a tariff to be levied on all imported goods, but their exemption was secured by the vigorous action taken by Stanley Unwin through the International Publishers' Congress and by direct letters to the President and leading members of the Althing. Reykjavik was notable for two bookshops specializing in books in the English language and with Snaebjorn Jonsson, the proprietor of The English Bookshop, Unwin had a close association.

EXCHANGE CONTROL AND RESTRICTION OF CREDIT

It was, then, by currency difficulties, by falls in exchange rates and by exchange controls, rather than by import duties, that publishers found their export business constrained. In March 1933 South African booksellers sought in vain from British publishers, as their Australian colleagues had previously done, more favourable trade terms to counteract the recent fall in their exchange rate. In Germany and Italy international trade became more difficult as the Nazis and Fascists gained control; and as their aggression threatened neighbouring countries, the difficulties expanded. In October 1934 the Association was informed by leading

[1] In France, in lieu of the internal tax on business turnover, an import tax of 2% was in force in 1932.

Berlin booksellers that since the unsold Reichsmarks in the special account at the Bank of England, which had been set up under the recent Anglo-German Exchange Agreement, exceeded the limit of 5 million, their remittances must be suspended until the account showed a less unequal balance and that while these conditions continued it would be very difficult for German firms to remit to England. In the following December a memorandum from the Department of Overseas Trade reported that the Nazi Government was forming a League of Reich German Booksellers to replace the existing trade association and that the buying of books and the livelihood of booksellers were suffering acutely from the racial boycott in Germany. The Association was forced to advise its members to supply for cash only until in February 1937 Stanley Unwin, on his return from a visit to Leipzig, reported that orders from German booksellers accompanied by a 'Zahlungsbewilligung' (permission to make payment) could be supplied in the expectation that payment would follow with little delay and that, to uphold the standing of its book trade, the German Publishers Association had undertaken to deal energetically with any bookseller who stated that he had permission to pay, but failed to do so. The Italian market was given particular attention by the Department of Overseas Trade in a valuable memorandum of February 1932, dealing with the promotion of British books abroad and the varying nature of the exchange restrictions; and a suggestion which it advanced, that Messagerie Italiane (the Hachette of Italy) should be appointed sole agent for the advertising and travelling of British books in Italy, was communicated to members of the Association, but failed to gain sufficient support. As in Germany, the Italian shortage of sterling in 1935 necessitated supply for cash only, but in November the purchase of books was excluded from exchange restrictions.

The severity of the restrictions which the Governments of some European countries were compelled to impose upon the import of books and their effect upon the solvency of booksellers can be seen from two examples. In 1937 the amount made available by the National Bank of Roumania for the import of books, periodicals and publications of all kinds from the United Kingdom was £200 per month, until in November it was doubled;[1] and in October 1938 the Czechoslovak Association of booksellers and publishers requested a loan of £30,000 from our Association to enable its members to meet their obligations to British publishers. For the avoidance of bad debts in these difficult trading conditions, in September 1934 the Association adopted a suggestion of R. F. West (of Baillière, Tindall & Cox) that it should operate a reference service on the financial standing of

[1] By July 1939 Roumania's monthly purchase of books and periodicals was averaging £900.

foreign booksellers and as a beginning should take over the list which that firm had been maintaining for the use of medical publishers. Supplemented also, after Stanley Unwin's visit to Leipzig in 1937, by the credit list of the German Publishers Association, the service covered almost 1,000 firms by 1939 and was to bring into being the Association's Credit Committee.[1]

THE BRITISH COUNCIL

A new force in the circulation of British books abroad came with the establishment of the British Council in 1935. In 1932 the Foreign Office had decided to encourage the formation of circulating libraries under the auspices of local institutes in certain European capitals, and subsequently in South America and elsewhere, and had put forward a scheme,[2] which the Association recommended to its members, under which the Foreign Office would make an initial gift of books up to the value of £50, provided that publishers would supplement it with a gift of half the amount, at published prices, of the books selected from their lists. From this prelude the British Council followed, and on 19 March 1935 four representatives[3] of the Association attended a meeting at the Foreign Office, to learn in particular that the distribution of books abroad would be one of the principal means by which the newly formed Council would seek to promote closer cultural and educational relations with other countries. At its meeting two days later the Council of the Association recommended publishers to support the work of the British Council by supplying one review copy of any book free and subsequent copies at half price, and by subscriptions to its funds; and it further agreed to make a donation of 50 guineas from the Association. In the following May the Association was invited to nominate a representative to the governing body and its choice of Stanley Unwin was to give the British Council a protagonist ready to fight its battles and, as chairman of its Books and Periodicals Committee, pre-eminently equipped to advise upon the development of the foreign review service, the planning of foreign libraries and exhibitions, and the commissioning of series of books and pamphlets on British life and culture.[4] Publishers were to have no regrets about the

[1] The Medical Group of the Association was still a principal source for this reference service in the 1950s.
[2] *Members' Circular* IX, no. 18, 132.
[3] Stanley Unwin (President), Bertram Christian (Vice-President), W. G. Taylor (of J. M. Dent & Sons; Treasurer), Geoffrey Faber.
[4] Published by Longmans. For a more detailed account of the British Council's use of books see Unwin, *The Truth about a Publisher*, pp. 418 ff.

launching of the British Council except that it should have come three years earlier.

RETENTION OF MARKETS

Although publishers had the help of this new ally in the propagation of British books abroad, they still had to fight their own battles on the trade front: to retain customary British markets in a reasonable division of world markets with American publishers; to stop the infiltration of unauthorized American editions into those markets when retained; and against local piracies in the Near East, India, and the Far East. With a large and expanding home market American publishers had not been seriously export-conscious until the great depression at home made them look abroad. But to British publishers, with a smaller home market, the retention of the foreign markets which they had cultivated was vital. 'The Council' wrote the President, G. Wren Howard, in 1938, 'has been at pains to point out that unless publishers always insist on obtaining all their customary markets, particularly Australia and New Zealand, and on the continent of Europe, they are likely one day to lose them altogether.'[1]

The Council had indeed been at pains to remind members of the Association of the threat to their cultivation of the European market since 1931, when in July it unanimously agreed:

that with regard to books of English origin, the English publishers should unite to resist any further encroachment by the American publishers on their market, and that they should aim to obtain in their contracts all rights in the English language outside the United States of America—with the possible exception of Canada—and also apart from possible reproduction in the Tauchnitz edition not less than one year after publication. Secondly, that with regard to books of American origin, the exclusive market in Europe should be reserved to the English publisher.[2]

When the National Association of Book Publishers of New York countered with a resolution that the European and Far Eastern markets had been in the past, and should remain, open to both British and American publishers, the Council stated most emphatically in October that, as regards Europe at any rate, the reverse had been the accepted trade practice.

[1] *Report of the Council, 1937–1938.*
[2] *Members' Circular* IX, no. 7, 44.

CHEAP CONTINENTAL EDITIONS

The problem of the early publication of Tauchnitz editions had arisen in 1927, when several Continental booksellers drew the attention of the Association to the growing practice of allowing these cheap paper-covered editions to appear simultaneously with the original editions. Since the release of these rights was not infrequently retained by authors in their own hands or in the hands of their literary agent, the Society of Authors was consulted and a joint committee of the two bodies recommended that no cheap Continental edition should be issued or announced within twelve months of the original publication. The representatives of the Association had pressed for a three-year, or failing that a two-year, delay—to give time for the exploitation not only of the 7s. 6d. edition of a novel, but also of the 3s. 6d. cloth-bound cheap edition—but were unable to carry the representatives of the Society. When in May 1928 the Secretary of the Society confirmed that out of some 400 author members who had replied to the recommendation only 12 had opposed it, its acceptance in practice seemed assured. But by July 1930 the Paris branch of W. H. Smith & Sons was writing to the Association to report numerous recent exceptions and in December, at the suggestion of the Fiction Group, the Council decided to send a questionnaire to leading Continental booksellers, in the hope that the replies might convince where the 1928 recommendation had failed. The replies which came from fifteen cities, from Helsingfors to Biarritz, were summarized in a letter to the Society of Authors of June 1931, as follows:

1. There was an almost unanimous opinion that the sale both of English fiction and of Tauchnitz editions was increasing on the Continent.

2. In answer to the question, 'Which do you prefer to sell, Tauchnitz editions or the English 7s. 6d. editions followed by cheap editions at 3s. 6d. or under?'—only one bookseller replied 'Tauchnitz'.

3. The question what effect a Tauchnitz edition has on the 7s. 6d. and subsequent cheap editions was answered with practical unanimity that the Tauchnitz edition seriously affected sales of the English editions. 'Most depressing'; 'kills the sale'; 'not to the profit of the author'; 'ruinous', were phrases used in answering this question.

4. In answer to the question, 'At what period, if any, should authors arrange for Tauchnitz publication?' two booksellers replied, 'Never'. The replies of the remainder average out at sixteen months.

In July came a reply from the Secretary of the Society,[1] that his Committee took the view that, save in exceptional circumstances, a delay

---
[1] D. Kilham Roberts had become Secretary in 1930.

of six months was reasonable for novels and of twelve months for other works and that, although the Committee was prepared to make a recommendation to that effect, the decision must rest with individual authors, with no prospect of unanimity in theory or practice. To that the Council of the Association retorted tartly: 'It appears that the almost unanimous evidence of Continental booksellers, so far from persuading your Society to reinforce their previous recommendation on the subject, has had the opposite effect of causing them very considerably to restrict their recommendation.'[1] It was therefore with evident relief that the Council received in December 1931 an assurance from the Society that on the division of markets between British and American publishers it would give 'every possible assistance in the matter if American firms show signs of attempting to put their Resolution into effect'; and the Council's minute closed with the following note: 'It was felt that this instance of the Authors' Society finding itself in complete agreement with the Association ought to be recorded.'[2] In April 1932 the Society took up again the problem of regulating the publication of cheap Continental editions and appointed a sub-committee to investigate it. When that Committee found itself unable to reach any unanimity, the Association was given the opportunity of stating its views against premature release of these rights in an article in *The Author*.

DOMINION MARKETS

Even more vital was the struggle, which came to the surface for the first time in the 1930s, to retain the British publisher's place in the Dominion markets. Of these Canada was inevitably the most vulnerable, and in negotiating contracts even with British authors American publishers could advance strong reasons of contiguity and cultivation for the Canadian market being thrown in with the American. Of the undesirability of Canadian book-buyers having to buy books by British authors through American publishers the Association was urgently reminded by the Department of Overseas Trade in October 1935 and although the Council urged members to fight for the retention of Canada in their agreements with authors, it begged the Department to use its influence also on the Society of Authors and the principal literary agents.

Throughout 1935 the Association was sending across the Atlantic a succession of protests against the supply to South Africa and India of

---

[1] The Society was requested to print the correspondence in *The Author*; and it was printed also by the Association (*Members' Circular* IX, no. 8, 54–7).
[2] *Members' Circular* IX, no. 11, 78.

British copyright titles in such reprint series as the Modern Library, Blue Ribbon Books, and the Garden City Dollar editions. But it was principally over Australia that the battle was fought: to retain the Dominion within British publishers' markets for their editions of books by American authors no less than by British and, when that had been won, to stop the infiltration of American editions. In November 1935 the Council recommended to all members a resolution of the Fiction Group advising them not to undertake the British publication of any book of American origin if the Australian market was excluded, and not to re-sell the Australian rights thus acquired to an Australian publisher, unless under very exceptional circumstances, and indeed emphasizing that the sale of rights in any work of fiction to an Australian publisher tended to weaken a central British control of the Empire markets. But the attack did not let up and at the monthly meeting of the Publishers' Circle in July 1937, after a discussion on the subject 'Are British publishers to lose Australian rights?', the Council was urged to take action 'to ensure that Australia and New Zealand may remain, as hitherto, the sole market of British publishers'. The resolution was referred to the Trade Committee,[1] to which Walter Harrap with his particular interest in Australia was added. Of the recommendations of the Committee the most important was that the Council should try to obtain members' signatures to an agreement under which they would bind themselves not to enter into a contract for any book from which the Australian and New Zealand markets were excluded, with the exception only of books by authors resident in either Dominion. The recommendation was adopted and by October 1937 seventy-one out of the ninety members had expressed provisional willingness to sign. Without unanimity some of the provisional signatories hesitated to confirm and an effective agreement had not been secured before the outbreak of war in 1939.

LOCAL PIRACY

The grip of the great depression upon international traffic in books inevitably increased local tendencies towards piracy and the Council became much exercised by its prevalence in Japan, the Argentine, Egypt and India. In Japan help was sought in 1931 from H.M. Consul. He recommended the appointment of a lawyer in Tokyo and himself intervened to secure the withdrawal of a piratical publication, but he reminded the Council that although a book originating from any country signatory

---

[1] The Officers (G. Wren Howard, W. G. Taylor, Geoffrey Faber), W. Longman, Daniel Macmillan, Sir Humphrey Milford, R. F. West.

to the Berne Convention was automatically protected against translation in Japan for ten years, confusion could arise from the existence since 1906 of a separate copyright convention between Japan and the United States which allowed, without further authorization, the publication in one country of translations of books originating in the other. The existence of this convention not infrequently led Japanese publishers into making translations from American editions of British authors under the impression that they were American publications which they were entitled to translate without permission. A warning of the growth of piracy in the Argentine was given in 1933 by the Camara Oficial del Libro in Barcelona and after the Tenth International Publishers' Congress in the summer of that year representations which all the constituent national associations made through their own Governments were followed by an improved copyright act. In Egypt also it was believed that a draft copyright act providing for adherence to the Berne Convention was in existence in 1934 and through the International Congress Stanley Unwin was active in organizing pressure for its implementation. In the same year the adoption by the Ministry of Education in Cairo of a text-book for school use was followed by the issue of a pirated edition by an Egyptian publisher; and to the Association's protest the Ministry replied that it could not insist upon the authorized edition being used outside its own schools.

THE INDIAN GROUP

It was in India, with its thirst for certificates of education and its low printing costs, that piracy was inevitably most widespread. In November 1933 the printing of a piratical edition of an anthology for use in schools led the Association and the Society of Authors to send a joint letter of protest to *The Times of India* and leading provincial papers; in April 1934 a protest was made to the India Office against an announcement by the Director of Education in one of the native States that he proposed to print school books and to make free use of any books he chose for translation purposes; and in the following July these and other examples led the Council on the proposition of Sir Frederick Macmillan, to appoint a Committee[1] to report on '(i) the growth of piracy of copyright material in India; (ii) the growing practice of educational institutions in India publishing text-books composed of copyright material based upon permissions from English publishers'. The principal recommendation of the committee was that members with interests in India should form

[1] R. C. Goffin (Oxford University Press), R. J. L. Kingsford (Cambridge University Press), W. Longman, Daniel Macmillan, A. E. Pigott (J. M. Dent & Sons).

themselves into a new Group; and in January 1935 the Indian Group held its first meeting, with Daniel Macmillan in the chair. It was agreed to appoint solicitors in Bombay, with powers of attorney to take action against piracy, to request all members of the Association and the Society of Authors to notify the Secretary of all copyright permissions granted or refused, so that a register could be kept, and to ask two Members of Parliament, Harold Macmillan and John Buchan, to emphasize during the Committee stage of the India Bill the importance of British India continuing to be, and the whole of the Indian Federation becoming, a signatory to the Berne Copyright Convention. It was subsequently agreed to form a fighting fund with which, if necessary, to contest cases of piracy in the Indian Courts, and to ask the managers of the Indian branches of British houses to form an Advisory Committee to watch for cases of piracy, to take action through the Customs to stop the entry of American Blue Ribbon and Star Dollar reprints of British authors, and to advise members upon applications from Indian Universities and native publishers for permission to use copyright material, particularly in anthologies to be prescribed for school-leaving examinations. By March 1937 the President, W. G. Taylor, was able to say in the Annual Report of the Council that all these activities of the Group and of its energetic Advisory Committee in India[1] had resulted in an appreciable diminution of piracies. An outstanding problem remained the risk of bad debts and in 1937 the Association began the circulation of six-monthly reports on Indian accounts.

COPYRIGHT IN TYPOGRAPHY

If the Association's main concern was to help its members to retain the foreign markets to which they were entitled under international copyright and by their agreements with their authors, the Council sought also to make international copyright itself more secure. Amendments to the Berne Convention approved in Rome in 1928, which came into force in 1931, were concerned mainly with reproduction by mechanical means and were not then of primary importance to book publishing. After the conclusion in 1932 of an understanding with the American Association of Book Publishers about the photographic reprinting in one country of books composed in the other, to which I have referred,[2] the Council took up in 1934 the question of copyright in typography. The Industrial Property Department of the Board of Trade was urged to press, in a revision of the Rome Convention, for its extension to give protection against the

[1] The first Chairman was F. E. Francis (Macmillan) with A. W. Barker (Longmans) as secretary.   [2] see p. 81.

unauthorized copying of typographical design and arrangement. The hope which was implicit in this proposal, that the United States might soon become a signatory to the Convention, was still to be hope deferred: in 1935 a new American Copyright Bill failed to get enacted. At the end of 1937 the Board of Trade was also urged to use the opportunity of the negotiations for an Anglo-American trade agreement to secure the abolition of the American import duty on British books. Although abolition could not be achieved, the agreement cut the duty on books of foreign authorship from 15% to 7½%, with effect from 1 January 1939.

BROADCASTING AND OTHER SUBSIDIARY RIGHTS

The growth of broadcasting and the coming of television drew attention to the value of rights of mechanical reproduction and so to other subsidiary rights. A new scale of minimum payments for the use of prose extracts from copyright works was agreed with the B.B.C. in February 1936, when a distinction was drawn between entertainment and other uses and it was conceded that the reading, for other than entertainment, of a passage of less than 200 words fell within the 'reasonable use' permitted by the Copyright Act; and in the following November a temporary arrangement was made with the B.B.C. to cover the use of a poem on television. The control of broadcasting rights provoked yet another disagreement with the Society of Authors. In October 1935 the Agreements Committee of the Association, which had been requested to undertake a revision of the model forms of publishing agreements, made an interim recommendation that in no circumstances should a member sign an agreement which permitted broadcasting of material from the work without his permission, that it was highly desirable that the publisher should have the sole right to authorize such use and that in that event at least two-thirds, and preferably three-quarters, of the receipts should be paid to the author. To this the Society of Authors' Committee of Management retorted that it was 'unanimous and emphatic in taking the view that any encroachment by publishers on authors' broadcasting rights should be resolutely resisted and that save in the most exceptional circumstances publishers should not expect to participate in the proceeds accruing from the exploitation of these rights by the author'.[1] The most which the Society could offer was to recommend its members that they should undertake, in cases in which they were asked to do so by their publishers, not to authorize the broadcasting of their books, or of extracts from them, at any time between the signing of the publishing agreement and six months following publication,

[1] *Members' Circular* XII, no. 1 (January 1936), 5.

provided that publication was not delayed more than four months from the date of delivery of the completed MS.

The Association and the Society continued to hold their opposing points of view, but in December 1936 Stanley Unwin reported to the Council of the Association that unofficial correspondence which he had had with the Secretary of the Society had led to the possibility of an agreement between the two bodies on subsidiary rights in general, and as a beginning it was agreed in the following May that a standing joint committee should be set up to negotiate with the B.B.C. on matters of common interest. W. G. Taylor, then Vice-President, Geoffrey Faber and R. F. West were nominated, with Stanley Unwin at call, to represent the Association and by January 1938 a draft agreement had been reached with the B.B.C., for signature by the three bodies, covering the use of copyright literary material in the Home and Empire programmes. The agreement, which came into force on 1 March 1938, provided for increased fees as compared with those in the previous agreements between the B.B.C. and the Society of Authors alone;[1] and the B.B.C., which had previously maintained that only when a payment was not made could the source and the publisher be acknowledged consented to make acknowledgment 'wherever the nature of the programme makes it practicable'. Subsequently in February 1939 the B.B.C. expressed the opinion that the parties were not *ad idem* in the interpretation of this condition and proposed that the agreement as a whole should be denounced and that an attempt should be made to settle terms which would leave no room for doubt. In discussion which then followed between the President, Geoffrey Faber, and officials of the B.B.C. the latter proposed 'that acknowledgment shall be made of the titles, authors and publishers of all copyright material included in programmes provided such acknowledgment is considered by the British Broadcasting Corporation to be of general service to listeners', and on the President's recommendation the Council adopted the new clause provided that, in applying for each permission, the copyright department of the B.B.C. would state whether full acknowledgment would be made, or whether acknowledgment would be made of title and author only, or whether no acknowledgment at all would be made. But the Society of Authors found itself unable to accept the new clause on the ground that it was less favourable to authors than the previous one; and the two bodies were still not of one mind when greater problems overtook them in September 1939.

Relations with the Australian Broadcasting Commission got off to an

[1] see p. 85.

uneasy start when in August 1937 the Commission sought free permission from a number of publishers to use extracts from text-books in its pamphlets for schools and subsequently complained of a lack of sympathy by the Association when its approach, which it regarded as a matter of courtesy and not of copyright, was not favoured. But when the attention of the general manager of the Commission was drawn to the use of some 17,000 words from text-books, mainly by one author, published by George Bell & Sons, a handsome apology and fee were forthcoming, and the storm was over. By 1939 agreements for the use of copyright extracts in broadcasting similar to that with the B.B.C. had been almost concluded with the Australian Commission and with South Africa and negotiations had been opened with the authorities in Canada and New Zealand.

GUIDE TO ROYALTY AGREEMENTS

In the course of its revision of the model forms of publishing agreements the Agreements Committee issued in December 1935, with the approval of the Society of Authors and the principal literary agents, a definition of the disputed term 'second serial rights': as '(*a*) any serial issue following a first serial issue in any newspaper or periodical, (*b*) any serial issue *after* book publication whether or not previously serialized'. After his unofficial correspondence with the Society of Authors in December 1936 Stanley Unwin was able to submit to the Council in the following May a memorandum on subsidiary rights in general, which had been approved by the Society in principle, and the Agreements Committee was then asked to draft suitable clauses for submission to the Society. But the work of the Agreements Committee[1] did not stop at subsidiary rights; it reviewed all the rights, markets and obligations to be assumed by a publisher in his agreements with his authors, and when the Association issued in 1938 its first *Guide to Publishers' Royalty Agreements*, it laid down the principles (with specimen clauses and notes and alternatives to meet a variety of circumstances) which were necessary to protect publishers against ill-informed or unreasonable authors and agents and to protect authors against rapacious publishers. The bulk of the document was approved by the Society of Authors and little of it had been seriously contested by that body.[2] 'In many ways' wrote the President, G. Wren Howard, in March

[1] The members of the Agreements Committee during its preparation of the Guide were: Stanley Unwin, W. G. Taylor, G. Wren Howard, Geoffrey Faber, G. O. Anderson, Bertram Christian, R. P. Hodder-Williams (Hodder & Stoughton), Harold Raymond.

[2] It might have been otherwise if the Agreements Committee had followed the lead of the Editorial Committee of the American Association of Book Publishers which in 1932 had recommended the adoption of contracts with authors providing for the calculation of royalties on the

1939,[1] 'the Guide may be described as a charter for the membership of the Association. While it in no way marks an attempt to return to the bad old days of publishing, nor results from a combined effort to reduce the financial rewards to which authors are entitled from the sales of their books, it is not a compromise accepted in the face of overwhelming force. It should be regarded generally as a fair and reasonable statement of the terms on which members are prepared to conduct their dealings with authors. It is complete and precise and yet provides that necessary measure of elasticity which enables it to be used in widely differing circumstances. It may perhaps be regarded as one of the most important and most useful pieces of work yet carried out on behalf of the membership'. So it was found, and, with periodical revision, has since continued, to be.[2]

TERMINABLE LICENCES

Within the area of publishing agreements the Council decided in October 1938 to give a lead against the acceptance of 'terminable licences', that is to say publishing agreements on a royalty basis terminable by either party at the end of a stated period of years, usually five or seven, without the occurrence of any breach of its conditions by the other party. Such agreements were generally regarded as inequitable, particularly if they contained no provision for the author to refund any unearned portion of an advance or to compensate the publisher for stock, plates and blocks left on his hands, but pressure from certain literary agents made it difficult for some publishers to refuse them on all occasions. The Council accordingly circulated three questions to the 124 members: would they sign an agreement: (1) never to accept terminable licences; (2) never to accept terminable licences in contracts with authors for whom they had not previously published; (3) to insert in all contracts providing for terminable licences a clause stipulating that the author, as a condition of termination, should buy all stock, moulds, stereos and blocks at cost or some agreed proportion of cost. To the three questions there were 29 noes to question 1, 22 to question 2, and 11 to question 3. It was decided in December 1938 to attempt to secure an agreement on the basis of question 3, but in November 1939 there were still outstanding four firms and in December two firms without whose signature the agreement could not

actual receipts at trade prices instead of on published prices, on grounds of logicality and of securing some reduction in royalties.
[1] *Report of the Council, 1938–1939*, p. 7.
[2] A model form for the sale of translation rights was added in the 1945 revision.

have been effective, and in 1940 the impossibility of finding time for the resolution of these persisting differences of opinion caused the attempted agreement to be abandoned.

LIBEL AND OBSCENE LIBEL

In two matters in which author and publisher were joint sufferers—the laws of libel and obscene libel—there was hope of alleviation, but it was cut short by the war. At the Council meeting in April 1935 the President, W. G. Taylor, drew attention to the recent prosecution by the police of James Hanley's novel, *Boy*, and an *ad hoc* Committee was appointed[1] to discover the facts and to report upon them. For the facts and the concern which they aroused we cannot do better than quote from a letter which was written to the press by E. M. Forster, A. P. Herbert, A. A. Milne, J. B. Priestley and H. G. Wells:

> 1. The prosecution occurred nearly $3\frac{1}{2}$ years after *Boy* had been published and after it had been reprinted four times.
> 2. Proceedings were initiated at Bury in Lancashire, where the police called at a local library and took possession of copies.
> 3. Messrs. Boriswood (the publisher) were indicted at common law. They were advised to plead guilty, for technical reasons, and on 20 March 1935, fines were imposed by Mr. Justice Porter at Manchester Assizes, to the extent of £400.
> 4. Messrs. Boriswood have now withdrawn *Boy* from circulation, but we are advised that legally they are still liable to proceedings in respect of every copy which may have been sold before withdrawal.[2]

The prosecution of *Boy* was quickly followed by the seizure by the police of a medical book also published by Boriswood and of some fifty books in a Bristol circulating library.

After the first meeting of the Committee Taylor and Unwin in an interview with an official of the Home Office expressed the concern of the Association at the state of the law and the procedure followed under it and in 1936 a memorial was submitted to the Home Office setting out, with illustrative cases, the anomalies and pitfalls, of which the following were the principal: the absence of any notification to the publisher that his books had been seized; the inadmissibility of evidence rebutting the charge of obscenity and supporting the literary and artistic merit of the work (although in one case a magistrate had admitted evidence on the ground

---

[1] Bertram Christian (Chairman), Geoffrey Faber, G. Wren Howard, Harold Raymond, W. G. Taylor, Stanley Unwin.
[2] *The Bookseller*, May 1935.

that the book was not fiction); the absence of any limit to the prosecutions of one book either in time or in number (and the consequent desirability of all cases being remitted to London); the apparent inevitability that the offence would be aggravated by a plea of not guilty; and finally the judgment of a book, not by its effect as a whole, but by extracts and even by the use of one or two words in it.[1] The memorial was effective and, although the law remained unchanged, the Committee was satisfied before the end of 1937 that a confidential circular issued by the Home Office to Chief Constables was likely to alleviate publishers' more serious grievances.

In July 1936 libel law reform was added to the Committee's terms of reference and it invited the co-operation of the Society of Authors, the British Federation of Master Printers, and the Empire Press Union. In 1927 Lord Gorell had unsuccessfully introduced in the House of Lords a Bill to give protection against speculative and 'blackmailing' actions arising from coincidental and unconscious libel and when representatives of the four bodies met in conference in 1936 it was found that, while authors and publishers, with their interest in works of fiction, were still particularly concerned with this risk, the concern of the press, and indeed the purpose of a Bill already drafted for the Empire Press Union, was only to provide that, with certain exceptions, no action for libel could lie without proof of actual damage. Although the Empire Press Union redrafted its Bill to meet these divergent interests it was felt by the Society of Authors and the Association's Committee that the result was not wholly satisfactory and in 1938 the Committee decided to make a fresh start and persuaded A. P. Herbert to promote in the House of Commons a Libel Reform Bill of his own. In the summer of that year the Committee was heartened by the case of Canning *v.* William Collins Sons & Co. in which the Lord Chief Justice gave a verdict against the plaintiff, who had identified himself with one of the characters in a novel although neither author nor publisher knew of his existence. When in November the Committee heard that A. P. Herbert had agreed to incorporate in his Bill certain principles from the Empire Press Union's Bill, in particular to give more latitude to newspapers and to provide that no damages could be awarded unless damage had been suffered, it feared that the resulting complication reduced its chance of success. Nevertheless the Bill went forward and when in February 1939 its second reading had been moved, the Attorney-General announced that the Lord Chancellor would set up a committee to consider the working of the present law, which he conceded

---

[1] This standard of judgment stood until 1954 when it was reversed by Mr Justice Stalde in *The Philanderer* case.

'did need overhaul and reconstruction'; and the Bill was thereupon withdrawn. The sittings of the Lord Chancellor's Committee on the Law of Defamation were suspended in 1940.

THE INTERNATIONAL CONGRESS

It was not until 1931 that the International Congress of Publishers regained its full status. The preliminary conference in 1929 at which, as we have seen,[1] the Association had been represented by W. G. Taylor and Stanley Unwin, had agreed to the constitution of a permanent commission of representatives of the national associations and to the re-establishment of the Bureau in Berne; and Unwin had been elected one of the Vice-Presidents. In 1931 the ninth Congress, the first since 1913, was held in Paris, and it was followed by the tenth in Brussels in 1933 and in 1936 by the eleventh at which our Association was once more the host in London.

The London Congress was held under the Presidency of Stanley Unwin, with W. G. Taylor as Chairman of the organizing committee. Entertainment included a reception at Lancaster House—the first recognition of the publishing trade by the Government—and banquets given by *The Times* and the Worshipful Company of Goldsmiths; and members of the International Commission were received in audience by King Edward VIII at Buckingham Palace. Readers of papers included Unwin on the right attitude of publishers towards translation rights, Faber on the exploitation of books by mechanical means such as broadcasting and talking machines and the importance of publishers retaining control of these uses, and R. F. West on the arguments in favour of book agents, which were then a matter of some controversy in the British trade.[2] Both in session and out the Congress worked hard and successfully for cooperation in the cause of books and to promote international goodwill. The final session, however, was touched by the shadow of Nazi Germany. When the head of the German delegation invited the Congress to meet in Leipzig in 1938, there was some opposition as a protest against the restricted liberty of the press in his country.[3] A proposal by an American

---

[1] see p. 79. [2] see p. 152.
[3] Sir Stanley Unwin recalled: 'Few of us wanted to go to a Nazi Germany, and an alternative invitation from Switzerland had been secured. Unfortunately my predecessor in office, M. Zech-Levie of Belgium, whose responsibility it was, had not even dropped a hint to the Germans of the possibility of another country being interposed, and without such an intimation they were fully entitled to expect us to honour our agreement to follow chronologically the order of the pre-1914 Congresses. Accordingly the German delegation came fully empowered (no doubt by Goebbels) to invite us to Leipzig in 1938, and were both surprised and horrified when the proposal was put forward that we should defer going to Germany and accept an invitation to Switzerland. Nazi pressure was, I believe, brought to bear on the Swiss Govern-

delegate that the invitation should be referred to the International Commission was defeated by 49 votes to 15 and on a second count the invitation was then accepted by 76 votes to one.

THE WEIMAR RESOLUTION AND FREEDOM OF PUBLICATION

That the shadow of Nazi censorship could cover the international circulation of books was seen in November 1936 when the Council of the Association received the following telegram sent from Weimar on behalf of the German book trade:

In the present chaos in the world unscrupulous instigators are endeavouring to drive Europe into a course of developments which must inevitably lead to catastrophe and consequently to jeopardising Western culture. While those conscious of their responsibility in all nations are aiming at peace, every art of lying and distortion is applied in order to instigate the peoples of Europe against each other. In this fateful hour of the West the representatives of all German publishers and booksellers gathered together in Goethe's city of Weimar on the occasion of the German Book Week mindful of the great responsibility which they bear as the transmitters of imperishable treasures of the mind undertook not to publish and circulate any books which with malicious distortion of historical truth insult the Head of a country or a nation or hold up to contempt the institutions and traditions which are sacred to a people. In the certainty that they hereby render a service to European peace they declare themselves willing to enter into an exchange of views with foreign publishers and booksellers who are prompted by the same imperative feeling of responsibility with a view to an international agreement.

It was not surprising that when the Leipzig Conference drew near members began to doubt the rightness of attending it and several American publishers declared a boycott. Stanley Unwin obtained an assurance from the incoming President of the Congress, Karl Baur, that there would be no restriction whatever upon freedom of discussion and that the so-called 'Weimar Resolution' would not be presented in any shape or form. For Unwin, the outgoing President, and Taylor, the other British member of the Commission, attendance was inevitable.

Freedom of publication was one of the issues for which the 1939–45 war was fought and it may be not inappropriate to end this chapter with words which appeared anonymously[1] in *The Bookseller* of 31 August 1939:

ment; anyway the Swiss invitation was withdrawn and the chief Swiss delegate made an impassioned speech in favour of acceptance of the German invitation.' *The Truth about a Publisher*, p. 405.

[1] The writer was John Hadfield, then with J. M. Dent & Sons, Director of the National Book League 1944–50.

One clear issue has emerged from the events of last week, however, and if I may be permitted to introduce a solemn note into these grotesquely comic proceedings I will briefly state this issue, for it vitally concerns our trade, and it will remain a real issue whether bloody war breaks out or the white war of nerves continues. It is one of the few issues that have actually gained in clarity as a consequence of the Russo-German pact.

It is simply, the issue of Light *versus* Darkness. That sounds absurdly grandiloquent; but it happens to be a fair representation of the case. Heaven and history know that a gallimaufry of threats, rivalries, ill treaties, bad economies, bluff and racial conflicts underlies the present struggle. In the present disposition of forces, however, the dividing line happens to have fallen by a seemingly automatic accident of history between those countries where there is freedom of expression and those where it is withheld.

As bookmen, a very simple issue confronts us. Let us leave democracy and totalitarianism out of the discussion for the moment. There is no present need to consider how Fascism differs from Communism (if at all), or how the doctrine of the sovereign state conflicts with the brotherhood of man. What we, as traders in literature, as authors, publishers, and booksellers, have to recognize is that those governments which appear to be aligned against Poland, or of which we have cause to be suspicious, are without exception governments which suppress, control, distort or abuse the printed word.

In the coming struggle whatever form it may take, we are likely to sacrifice much of our freedom of expression. That is one of war's little ironies. Let us not forget this cardinal issue, however. Freedom of expression in the printed word is the very essence and justification of our trade. Let us keep it before us as a symbol of what we are struggling for even if it shines no more brightly than a blue lamp in a black-out.

8

1931-1939 (2)
Book clubs, Book Tokens, book weeks

If, as was true in the early 1930s, retail bookselling outside such exceptional localities as the University towns offered a precarious livelihood unless supported by the more profitable sale of stationery and fancy goods, it is not surprising that many booksellers saw 'a third off' as the only cure and an infallible one. New ideas and new channels by which the number of book-readers and book-buyers might be increased were often seen as a threat to today's book-selling rather than as giving promise of an increased number of book-buyers tomorrow. Nevertheless these were years of inventiveness and experiment, in which progressive minds on both sides of the trade tried to overcome the effects of the economic blizzard and, inspired sometimes by a spirit of social evangelism, to find means of bringing books to the 'fringe public' which was diffident at entering a bookshop and often unable to afford the general run of book prices. These were the years of the 3*s.* 6*d.* cloth-bound reprint series—the Travellers, the Phoenix, the Windmill, the Adelphi Libraries; of Benn's Sixpennies; of the pioneer work of the Phoenix Co., inspired by John Baker with the backing of Hugh Dent,[1] in the sale of books on the instalment plan; and in 1935 Allen Lane produced his first Penguins. Nor should Victor Gollancz's attempt to encourage readers of novels to buy, rather than to borrow, go unremembered. In 1930 he formed a new company, Mundanus, with the announced aim of publishing initially at the rate of one a month, and at the price of 3*s.* net, new full-length novels in paper covers.[2] Three were issued in 1930, but the unwillingness of booksellers to put books in soft covers on to their shelves compelled the abandonment of the experiment in the following year. New novels did not get the support of the newsagents which enabled the first Penguin reprints five years later to survive the initial opposition from booksellers.

---

[1] The son of the founder of J. M. Dent & Sons.
[2] The design of the yellow paper covers surely shows the hand of Stanley Morison.

BOOK CLUBS

These years brought also the beginning of the book club movement. Its precursor, The Book Society, had been formed in 1929[1] and with Hugh Walpole, Clemence Dane, George Gordon, Sylvia Lynd and J. B. Priestley as its first selection committee undertook to deliver to its members a monthly choice on the day of publication. To the booksellers assembled at Hastings for their annual conference that summer David Roy,[2] a forward-looking leader of the trade, saw this new development only as a threat of decreasing business. 'Doubtless every trade' he said,

has its own trouble, and our particular misfortune seems to be that there are always a large number of people ready to rob us of our 'easy' business—that section of our turnover which is compensation for all the non-profit-making business we must necessarily do. The drapery stores oblige us in this neighbourly way at Christmas time when they go all out for the annuals and reward business, but make no effort to compete with us during all the other months of the year in the carrying of a general stock...These clubs come into the same category. Not for them the daily worry and vexation of stock buying and stockkeeping or the handling of customers' profitless orders. One order for 3,000, 5,000, 10,000, 15,000 books and all of them sold in advance. This is the book trade as they see it.

Following Roy's lead the conference passed a resolution expressing the hope that publishers and authors would set their faces against the activities of the Book Society. But the Book Society was doing no more than supply to its members a chosen book at the ordinary published price and was in effect a mail-order bookseller conforming to the Net Book Agreement; and the same was true of the Junior Book Club, which was recognized for supply by publishers at ordinary trade terms in January 1933.

The idea of book clubs caught on and the ability to place one order for 3,000, 5,000, 10,000, 15,000 or more books, all of them sold in advance, meant for the book club low distribution costs and for the supplying publisher a considerable reduction in his manufacturing cost. There were soon four political clubs (Left, Right and centre), two religious clubs, and two general (The Readers Union and Foyle's The Book Club), all eight supplying their members at prices substantially lower than the ordinary published price, usually at 2s. 6d. Gollancz was first in the field with his Left Book Club in 1936 and with the support of meetings in the

---

[1] Its founder, Alan Bott, was thus the introducer of the book club movement to this country from the United States.
[2] of W. H. Smith & Sons.

Albert Hall and in Hyde Park the number of its members rose to approximately 60,000. It was followed in 1937 by the first of the general clubs, John Baker's Readers Union, offering 15s. and 10s. books at 2s. 6d. and having within a year an enrolment of 17,000 members, and in the same year by the Student Christian Movement's Religious Book Club, offering its choices at 2s. It was not long before publishers as well as booksellers began to fear the effect which two widely different prices for the same book might have on the mind of the public.

In the summer of 1937 the Associated Booksellers, assembled in Cheltenham for their annual conference, passed a resolution embodying principles by which book clubs should be regulated. They proposed that club organizers should endeavour to work within the existing retail system and should so word the coupons attached to publicity as to direct enrolments through booksellers; that book club prices should leave room for a reasonable profit for booksellers and should not be post-free; and that any publisher who made an arrangement with a book club for the issue of a book at a cheaper price than the original without having announced it at the time of the original subscription to the trade should be prepared to consider any stock of the book held by a bookseller as having been supplied 'on sale or return'. The resolution, which was passed on to members without comment by the Council of the Association, was followed in September by a strong protest from the Booksellers' Council against Foyle's announcement that H. G. Wells's novel, *Brynhild*, was to be supplied to members of their newly formed Book Club at 2s. 6d. immediately after its general publication at 7s. 6d. The Council of the Association was at last moved to take action and the September issue of the *Members' Circular* contained the following resolution set in heavy type:

that in view of the effect upon the public's attitude towards book prices, the Council is of opinion that as a general principle it is most unwise and contrary to the interests of the Book Trade for publishers to authorize the publication of their books, certainly those of a general character, in special book club editions at a reduced price within a period of at least six months after first publication.

This was not a full reply to the resolution from the Booksellers' summer conference and at its November meeting the Publishers' Council received a memorandum suggesting rules for the regularization and administration of book clubs including their division into two categories—those having a *bona fide* propagandist purpose and those existing for other purposes—and proposing the appointment of a Joint Committee. The Council agreed that the proposals should be considered and nominated W. G. Taylor, Geoffrey Faber and Daniel Macmillan to represent the Association on a joint Committee, to which the Associated Booksellers appointed David Roy, who became chairman, Basil Blackwell and R. G. Davies. The committee was subsequently enlarged by the addition of G. Wren Howard and Walter Harrap and of three additional booksellers, A. S. Jackson, J. H. Ruddock and C. W. Cragg.

In January 1938 the Committee submitted an interim recommendation: that it was desirable that book clubs should operate within the net book system and that all existing clubs and any new ones should be asked to register with the Publishers Association their rules and conditions of membership and the titles of their selections not less than one month prior to issue and to supply only their registered members. A questionnaire to the existing clubs elicited no objections and a first report of the committee was issued in April. The main recommendation was to the effect that if the price of the trade edition was not more than 40% above that of the book club edition, the appearance of the two editions need not be separated by any interval of time. Otherwise an interval of at least 12 months must separate the two editions, whichever came first. The 40% ratio was that between 3s. 6d. and 2s. 6d. At that time the consideration uppermost in the minds of both publisher and bookseller members of the committee was the danger to which the trade would be exposed by any wide disparity between the price of a club edition and that of a trade edition appearing simultaneously or within a few months of each other. The publisher members, in particular, felt that such a disparity would be taken by the public at large as confirming the common, if fallacious, view that books are needlessly expensive; the bookseller members, sharing this apprehension, objected to a state of affairs which would expose them to the complaints and criticisms of their customers. The report was adopted by the Councils of the two Associations and the Committee was instructed to work out the regulations for its enforcement. But objections then began to arise from the clubs, voiced in particular by Victor Gollancz on behalf of his Left Book Club in a meeting with the publisher members of the Committee, and from the four principal circulating libraries, who feared that the advantages of club membership would reduce the numbers of their subscribers and the sale of their *ex libris* copies. Although the bookseller members of the Committee were inclined to enforce the regulations upon Gollancz as they stood, to the publisher members it was inconceivable that the Publishers Association could embark on a course of action against one of its members which might result in extensive black-listing of booksellers and even of publishers. In March 1939 the further deliberations of the Committee, now enlarged in personnel, issued in a second

Book clubs 137

report, with accompanying regulations, in which the principal difference was that the price ratio of the trade to the club edition, if both appeared simultaneously or within twelve months of each other, was altered from 40% more to three times more. In other words, if the club edition was priced at 2s. 6d. the trade edition might be priced at 7s. 6d. but not more. This revised code was adopted, though with reluctance, by the Council of the Associated Booksellers as a compromise satisfactory only because it had been accepted by all parties. By the Council of the Publishers Association it was referred to a special General Meeting in April.

At the meeting alarm was expressed at the harm which would be done to the trade if novels and other 7s. 6d. works of general literature were to be issued in half-a-crown club editions within twelve months of publication and by a resolution passed *nem. con.* the report was referred back to the Committee 'for reconsideration of the regulations contained therein, in the light of the proviso that no book, other than those issued by political or religious book clubs, should be issued in a book club edition at a reduced price within twelve months of the date of publication of the book in a trade edition'. With this clear lead the Committee found no great difficulty in producing in May its third and final report and regulations, and they came into force on 1 June 1939. The ratio of 7s. 6d. to 2s. 6d. was retained for books issued within twelve months of the trade edition (whether before or after) by a political or religious book club *recognized as such*; for all other book club issues the principle was laid down that they must have been published previously in a trade edition and that the interval must be at least twelve months. The regulations also covered the conditions of club membership, the timing of announcements of forthcoming club choices, and the circumstances in which booksellers and circulating libraries could apply for the allowance of credit on stocks of trade editions held by them. The Committee was entitled to defend itself against criticisms aroused by the varied recommendations in its successive reports. Widely divergent views were held by promoters of book clubs and by publishers and booksellers and, in the words of the final report,

the task of bringing these views to a common point is probably the most difficult ever essayed by a book-trade committee...During the period of the committee's existence not only has the book club movement developed and ramified and the personnel of the committee undergone change, but no clear indication of the general will has been given until the Publishers Association adopted the resolution reproduced above...a phrase in that resolution ('issued by political and religious book *clubs*') has suggested the solution of a problem which had earlier defeated the committee. In its first meetings the committee had considered the possibility of making a distinction between propagandist *books* and other kinds of books, and was finally forced to abandon this line of approach to the problem because of the impossiblity of defining the terms used. The solution now adopted seems simple and even obvious. Something similar might be said of many inventions which had to wait a long time before anybody thought of them.

The joint Committee, reduced in size, remained in being until the end of 1941 to administer and occasionally to interpret the regulations, to inspect registers of club membership and to vet new clubs. The recognition of the clubs as political or religious was kept by the Councils of the two Associations in their own hands, and those initially recognized were: The Left Book Club, The Right Book Club, The National Book Association, The Liberal Book Club, The Peace Book Club, The Religious Book Club (Foyle), The Religious Book Club (S.C.M. Press), The Catholic Book Club. In 1940 a new clause was added to the Decisions and Interpretations relating to the Net Book Agreement: it pronounced it to be a breach of the standard conditions of sale for a bookseller to sell a book club edition to a member of the general public who was not registered with that bookseller as a member of the club, except as a second-hand book.

BOOK TOKENS

'If I give him a book perhaps I shall choose one which he has already read and even if not, do I really know what he likes?' The solution was provided in 1932 by the Book Token, the invention of a publisher, Harold Raymond. He has said that the idea had been in his mind for seven or eight years and in 1926 he addressed the Society of Bookmen on the subject.[1] In 1928 he put it to the National Book Council, which appointed a sub-committee to investigate it. It was decided that the N.B.C., not being a trading body, could not handle the scheme itself and that in any case it would be advisable to pass it over to the Associated Booksellers without whose official backing Tokens would be no more useful than postal orders which could be bought and cashed only from a few enlightened post offices. A statement was prepared for submission to the annual conference of the Associated Booksellers at Hastings in 1929, but it was crowded out by more urgent business, in particular the negotiation of the agreement with the Library Association. The N.B.C. took up the scheme again in 1931 and in 1932 it was introduced at the Chester conference by Basil Blackwell, to whose wisdom and persuasiveness in negotiation much has been owed by publishers no less than by booksellers. The conference

[1] The address, under the title of *Selling Books by Coupon*, was printed in *The Bookseller* of 16 September 1926 and in *The Publishers' Circular* of 2 October 1926.

approved the scheme in principle and instructed its Propaganda Committee to explore its practicability with the assistance of Harold Raymond and Maurice Marston, the Secretary of the N.B.C., and the Propaganda Committee in its turn recommended that the scheme should be instituted at once and to that effect appointed a Committee to administer it. That Committee met in July 1932 and decided to ask the National Book Council, in spite of its previous decision to the contrary, to manage the scheme on behalf of the Associated Booksellers, and Douglas Leighton,[1] Treasurer of the N.B.C., was accordingly invited to join the Committee. That first Book Tokens Committee, consisting of three booksellers,[2] the Treasurer and Secretary of the N.B.C. and Harold Raymond, was in effect a committee of the N.B.C. and all its meetings with the exception of the first were held in the N.B.C. offices. For three months the Committee was fully occupied with the details of the scheme and it was not until 19 October that its responsibility was defined in a letter to the Associated Booksellers. Of this letter, which was subsequently held to be legally binding, the Publishers Association was not officially informed and the exploitation of it was to be the cause of considerable and continuing resentment, in part reasonable, in part unreasonable, in the minds of prominent publishers: reasonable in its opposition to the appropriation by the Associated Booksellers to its own funds of an unexpected windfall of profits from a scheme which had been invented by a publisher for the benefit of authors, publishers and booksellers; unreasonable in that the Booksellers were legally the owners of the scheme.

In this letter of 19 October 1932 the Committee of the National Book Council set out the terms on which it undertook to administer the Book Tokens scheme for the Associated Booksellers. It admitted that 'the principle of the scheme is the property of your Association' and that 'all documents and stamps used in the scheme are, so far as my Committee are concerned, the copyright of your Association'; and it undertook to account to the Associated Booksellers for any net profits earned and in their calculation to charge only for extra expenses incurred in the operation of the scheme, without any contribution to the general overheads of the N.B.C. A Book Tokens Committee of five was regularized, three members to be appointed by the Associated Booksellers, two by the N.B.C.[3] Finally, the letter relieved the N.B.C. of any financial liability in

[1] of Leighton-Straker Bookbinding Co.
[2] Basil Blackwell, Maurice Hockcliffe (Times Book Club), Hubert Wilson (London).
[3] The first members were: M. Hockcliffe, D. Leighton, Cadness Page (Harrods), Harold Raymond, Hubert Wilson.

connection with the scheme and implicitly put it on the shoulders of the Booksellers. It is important to recall that, although a loan of £100 was collected from trade bodies to enable the Committee to make a start, the Associated Booksellers shouldered the responsibility for the flotation expenses in the event of failure and that the operation of the scheme was thought in itself to be a service to the N.B.C. in giving it an additional *raison d'être* at a time when, owing to the slump, it seemed to be in danger of losing support.

The scheme was launched on 14 November 1932 and when the results of the Christmas season were known there was promise of great success provided that sufficient booksellers could be persuaded to give it proper support. It comprised a greetings card priced at 3*d*. with stamps of various denominations from 3*s*. 6*d*. upwards. On the sale of the card the bookseller received no discount; on the face value of the stamp there was a trade discount of $12\frac{1}{2}\%$ and the $87\frac{1}{2}\%$ which the issuing bookseller paid to the Book Tokens Committee was passed on to the redeeming bookseller, who being presumed to have bought the book from the publisher at a discount of 25% (on average) would also have a margin of $12\frac{1}{2}\%$. The difference between the manufacturing cost of the Token and the 3*d*. charged for it represented the only anticipated source from which the Committee could meet its postage, publicity and overhead expenses, and it was not assumed that any profit would be left over for the Associated Booksellers as owners. But there was to be an unexpected source of revenue: if a Token was not redeemed the Committee continued to hold $87\frac{1}{2}\%$ of its face value.

The first report and statement of account were submitted to the Bath Conference of the Associated Booksellers in the summer of 1933. Although the scheme had been accepted at Chester without dissension there was a large body of booksellers who were refusing to co-operate in it. Several important men were openly hostile and one whole Branch, the North Eastern, had recommended to its members a boycott. In spite of the recommendation of their Council and a personal appeal by their President less than 300 of the total of nearly 1,000 members of the Associated Booksellers came in at the beginning, and of the supporters some were less than half-hearted and could be heard saying that they would keep Tokens out of sight under the counter. It should, however, be recalled that the flotation of the scheme coincided with a trade row over cigarette coupons, and by one of those quirks of which the human mind is capable the two became confused. Early in 1933 there was no feeling of assurance that the scheme was safe and the expectation of profits accruing to their Association was dangled before the eyes of reluctant booksellers. A memorandum on the work of the Book Tokens Committee presented at a National Book Council luncheon on 7 April 1933 held out the hope that if all the members of the Associated Booksellers would co-operate 'a sale of anything over double the first year's figures will bring a very pretty contribution to the Association's funds'; and the report presented at the Bath Conference similarly referred to the hope of 'a substantial profit, which would strengthen the slender finances of our Association and perhaps even provide for further schemes for extending the sale of books'. That the Associated Booksellers were the legal owners of the scheme and that these inducements had been offered was insufficiently recognized when in subsequent years publishers protested that unredeemed Tokens represented a loss to authors and publishers no less than to booksellers and that any profits should be applied by the Associated Booksellers, not to strengthen their own finances, but as trustees to provide for further schemes to extend the sale of books.

As the account to which unredeemed Tokens were credited began to grow, the Book Tokens Committee decided that the proper use of it was in publicity for the scheme, but nevertheless early in 1934 it decided also that as the account was drawn on for that purpose $12\frac{1}{2}\%$—representing the amount which the booksellers would have gained if the Tokens had been presented—should be handed over to the Associated Booksellers. In the same year the free contribution by the National Book Council of some of the administrative cost was also reviewed and the Associated Booksellers agreed that £300 p.a. should be paid. By the summer of 1935 it appeared to the Committee that the Finance Committee of the Associated Booksellers, of which H. E. Alden was Chairman, were hoping to secure still further contributions out of the scheme and Alden suggested that a member of his Committee might attend meetings of the Tokens Committee with a watching brief over large items of expenditure. The Book Tokens Committee, believing that it could not satisfactorily perform its executive control under surveillance of that kind, decided that its rights and obligations must be defined in a new agreement replacing the informal letter of October 1932.

There followed long negotiations, difficult at first, between the Book Tokens Committee and the Finance Committee of the Associated Booksellers. The more extreme members on the booksellers' side claimed that their Association owned the scheme absolutely, that the idea of trusteeship was a new gloss on the part of the publishers and that they were under no obligation even to listen to advice from other trade bodies; but many prominent booksellers including those on the Tokens Committee saw beyond legal obligations and recognized the reasonable right of the other half of the trade to be heard. Much was owed to the large-minded intervention of Basil Blackwell as President of the Associated Booksellers and agreement was finally reached under his chairmanship. The constitution of a limited company was at first proposed by the Tokens Committee as a means of achieving executive power and of meeting the standing claim of the Associated Booksellers that the scheme involved them in possible financial liability, but the Finance Committee, to which such liability could no longer have seemed a real threat, rejected the proposal. Its abandonment was to provoke a somewhat dramatic resignation.

In November 1935 at the request of the Tokens Committee Geoffrey Faber, who was at that time chairman of the Executive Committee of the National Book Council, wrote a letter to Blackwell urging acceptance of the proposed limited liability company, in which the Associated Booksellers would retain its dominant position, and in December to fortify the Tokens Committee in its stand he sought and obtained the backing of the Council[1] of the Publishers Association, to some of whom at least knowledge of the absolute control given to the Associated Booksellers in 1932 came as a surprise. There is some doubt whether Faber verbally informed the Secretary of the Tokens Committee of this support for the incorporation of a company; and in his turn he was not informed that the Tokens Committee and the Finance Committee were jointly putting forward an alternative solution. The first he knew of it was when he took the chair at a meeting of the Executive Committee of the N.B.C. on 9 January 1936. Although detailed consideration of the proposed scheme and of the reasons why the Tokens Committee were prepared to recommend it was deferred to a special meeting, Faber took the view that it was a *fait accompli*—a wrong view, it seems, because the Tokens Committee had no plenary powers from the Executive Committee—and that he had put the Council of the Association in a false position. That evening he resigned the chairmanship of the National Book Council.

The scheme was approved by the Council of the Associated Booksellers and subsequently by the Executive Committee of the National Book Council, but the latter, aware that Faber had invoked the interest of the Publishers Association in December, decided also to communicate the scheme to the Association before taking further action. At its February meeting the Council passed the following resolution:

Since publishers are no less interested than booksellers in the efficient management of the Book Tokens Scheme, as in any other co-operative scheme for the

[1] The discussion is not minuted.

wider distribution and increased sale of books, this Council regrets that it was not consulted with respect to the Agreement of October 1932, and it invites the Associated Booksellers to consider whether, in the interests of harmonious and effective co-operation between the two Associations, the proposed new agreement should not now be the subject of consultation between the Publishers Association and the Associated Booksellers.

A second resolution appointed a committee consisting of the officers, W. G. Taylor, Stanley Unwin and G. Wren Howard, with R. F. West, who was a member of the Council of both Associations,[1] to examine the new agreement and the circumstances relating to its preparation.

The committee met four representatives of the Associated Booksellers, led by Blackwell and including David Roy, on 26 March 1936 and the minutes read as a tribute to the two Presidents, whose concern was not with charges of unjustifiable interference or rapaciousness but to restore and extend good relations between the two Associations. The scheme which had been recommended by the Tokens Committee and the Finance Committee of the Associated Booksellers was accepted with minor alterations, and the only threat of disunity came from the insistence of the Associated Booksellers' solicitor that they must have the right to dispense with the services of the National Book Council if their ownership of the scheme were not to be nullified. The agreement, which was signed in May, acknowledged that the primary object was not the raising of revenue for any one section of the trade and although it reaffirmed absolute ownership by the Associated Booksellers it abolished the payment to them of the $12\frac{1}{2}\%$ discount on unredeemed tokens and substituted for it a royalty of 10% on the annual sales of the $3d.$ card up to 160,000 cards and of $2\frac{1}{2}\%$ thereafter, with a current annual limit of £250. It further provided that while the N.B.C. would hand over on request any sums surplus to the Tokens Committee's requirements for the administration and development of the scheme, the Associated Booksellers would 'not be entitled to such sums otherwise than in connection with another scheme or schemes for the increase of the sale of books'. Payment to the N.B.C. for the administrative expenses was confirmed and it was laid down that one of the two N.B.C. members on the Tokens Committee would be a member of the Publishers Association and that, in the event of the scheme not being operated by the N.B.C. to the satisfaction of the Associated Booksellers, the latter would not terminate the agreement without reference to

[1] At this time Blackwell suggested that there should always be one member common to the Councils of both Associations, but it was not pursued.

an *ad hoc* committee consisting of three persons appointed by the Associated Booksellers and two appointed by the Publishers Association.

The agreement should have ended the resentment between the two sides of the trade of which Harold Raymond's happy invention of Book Tokens had so unnecessarily been the cause. But ignorance of the 1932 letter, which had made the Associated Booksellers owners and not trustees of the scheme, and of the inducements of fat profits which had been dangled before them in 1933 persisted in the minds of some publishers, and to some booksellers the intervention of the Publishers Association continued to seem unjustifiable interference in an agreement which concerned only their Association and the N.B.C. Again in 1938 the Publishers Association was invited by the N.B.C. to intervene when the Associated Booksellers had been asked to agree to the management fee being increased from £300 to £500 and had in return demanded that its own royalty on the sale of Token cards should be a flat 10% without limit. Of both these requests the appointed representatives of the Publishers Association—G. Wren Howard, Walter Harrap and F. V. Morley (of Faber and Faber)—advised acceptance.

It may be convenient here to continue the later history of the Book Tokens scheme in so far as it falls within the years covered by this history. In 1940 by agreement with the National Book Council the Associated Booksellers set up a trust fund to receive part of any surplus for the general benefit of the book trade; and when in 1942 income tax problems necessitated the formation of the scheme into a limited liability company, they proposed also the formation of a joint National Book Trade Association[1] which would be endowed with the Book Tokens scheme. But the composition and powers of the board of the company again caused disagreement between the two Associations. Even though the A.B. yielded to the P.A.'s request that each Association should have equal representation on the board, the proposed articles of association made it clear that the board was to be allowed no power of action in matters of policy. The three intended nominees—G. Wren Howard, Harold Raymond and W. G. Taylor—refused to serve and in 1943 the Council of the P.A. declined the invitation to appoint three publisher directors. There followed a period of over twenty years during which the Book Tokens scheme was operated without any relationship between the company and the P.A. or any individual publisher.[2]

[1] see p. 197.
[2] The happy ending which was to come falls outside the scope of this narrative. Under the wise chairmanship of H. L. Schollick (of Blackwell's) during the years 1943-68 the board of

GIFT COUPONS

It has been mentioned that the early antipathy which some booksellers felt towards Book Tokens arose from the confusion of gift tokens with gift coupons. In 1932 Messrs Wix & Sons were giving coupons with their Kensitas cigarettes and in October they issued a list of 450 books, issued by fifteen publishers, which could be obtained in exchange. Throughout November and December protests from booksellers filled many columns of *The Bookseller*. 'Here' wrote David Roy,

is the biggest departure from the recognized method of selling books that we have seen for a very long time...Booksellers had really begun to believe that they were looked upon by the publishers as joint partners in the business of book distribution. But their optimism has been dealt a very heavy blow, for here is a revolutionary scheme launched by some of the best houses in the trade, without the publishers concerned having even troubled to look over the fence and ask the Associated Booksellers what they thought of the idea.

Why, he asked with some perversity, should tobacconists be allowed to give a book to every purchaser of a guinea's worth of cigarettes when booksellers were prohibited from giving a packet of cigarettes to every purchaser of a guinea book? At its November meeting the Council of the Association received an official protest from the Associated Booksellers and several letters from its own members requesting consideration by the Association as a whole; and a special General Meeting was called.

At the meeting on 24 November 1932 members were confronted with three questions: whether coupon trading in general was desirable; whether, if the answer to the first question were in the affirmative, books should be included among the articles which could be acquired in exchange for coupons; and whether steps should be taken to ensure that net books should not be obtainable by coupons at less than the published price. On the first two questions there was no likelihood of unanimity. Sir Frederick Macmillan read letters from eleven prominent authors who

Book Tokens Ltd has done much for the common good of the book trade. Among these services the following deserve particular mention: annual donations and other financial help to the National Book League; support of National Library Week; a contribution of £5,000 to the cost of defending the Net Book Agreement before the Restrictive Practices Court in 1962; a contribution of £2,000 to the expenses of the World Book Fair in 1964; the flotation of Book Trade Improvements Ltd to provide finance for booksellers particularly for the improvement of their premises; an annual subscription of £1,000 to the National Book Trade Provident Society and the endowment of two bungalows at the Society's Retreat as a tribute to Harold Raymond, the inventor of Book Tokens. Since 1948 Book Tokens Ltd has also operated the Booksellers Clearing House for the payment of booksellers' accounts with publishers. The board of directors now includes publisher members.

had written to Messrs Wix in commendation of their scheme and he proposed a resolution whitewashing those publishers who were participating in it, but recommending that new contracts should ensure that net prices should be maintained in the calculation of coupon values. Amendments requiring members not to participate in any form of coupon scheme without first reference to the Council of the Association and recommending a nine-months' standstill, during which an assessment could be made of the effect of the existing contracts upon the sale of the particular books and upon bookselling generally, were not supported. Sir Frederick's proposal was carried by thirty-two votes to one and a committee of four—C. W. Chamberlain, William Longman, Sir Frederick Macmillan, W. G. Taylor—was appointed to give effect to it; but nevertheless it was also agreed to invite the Associated Booksellers to send a small committee to a joint discussion, with nominees of the Council, of the whole question of gift coupon schemes. The committee of four met the Managing Director of Messrs Wix on 29 November and they were able to inform the Council at its December meeting that the number of coupons required in exchange would be so increased that it would no longer be possible for the public to obtain, by means of Kensitas coupons, net books at less than the full published price. But in the joint discussion early in January the booksellers' representatives maintained their opposition to the scheme as derogatory to the standing of books and the Council yielded to it in suggesting that the publishers concerned in the scheme should meet at the earliest possible date to discuss any extension of it. They met on 27 January 1933 and agreed to give notice to Messrs Wix that they would not extend their existing contracts beyond twelve months and that they would not add any further titles to Wix's catalogue. The booksellers' objections had carried the day and it would be pointless to speculate whether in time the offer of books for coupons would have won new readers and have brought new buyers to bookshops; on 31 December 1933 not only did Messrs Wix cease to offer books, but by agreement within the Tobacco Trade Association the issue of gift coupons with cigarettes was discontinued.

The offer of books for cash and coupons was attractive also for a time to newspaper proprietors. In 1933 readers of the *Daily Mail* were offered a complete Shakespeare for six coupons and 5*s*. 9*d*., which was countered by Basil Blackwell's publication of the Shakespeare Head edition at 6*s*.; and in 1934 the *Daily Herald* offered a special edition of Bernard Shaw's forty-two plays for six coupons and 3*s*. 9*d*. Shaw acted against the advice of Constable's, who published for him only on a commission basis, and his action was bitterly resented by booksellers. In a statement to *The Bookseller* in October he asserted his conviction

that this particular transaction will increase the business of every bookseller in the country by adding to the book-reading public many thousands of customers to whom 12s. 6d. books are as inaccessible as Rolls-Royce cars. I must not, however, suggest that there is anything outside the ordinary and inevitable routine of business in my dealings with the Odhams Press. The field is open to all distributors, whether they operate on the scale of Odhams, Smith & Sons and Selfridge's, or on that of the most modest single bookshop. But an order for one, two or three single copies cannot be executed at the rate of an order for thousands of copies. If the Woolworth firm sends me an order of sufficient magnitude to enable it to sell copies at its stores for 6d. and yet give me a better return for my labour than I can obtain through prices that are prohibitive for nine-tenths of the population, I shall execute that order joyfully. And let the bookseller who would do otherwise cast the first stone at me.

COUPON ADVERTISING

If there had been some confusion between Book Tokens and gift coupons, there could have been even more between gift coupons and advertisement coupons. In February 1938 the Publishers' Advertising Circle[1] suggested to the Council of the Association that the success of Messrs Simon and Schuster's American experiments in direct selling by advertisements with order coupons indicated the desirability of a controlled trial in Britain and drew attention to the fact that Simon and Schuster's wide-scale advertising had not only created a large direct sale but had given a great impetus also to sales through bookshops. The Circle reported that Bernard Watson of Ivor Nicholson & Watson was ready to make an experiment with a suitable set of books if assured that it would not excite hostility from booksellers and it proposed that nominees of the two Associations should supervise the experiment and prepare a report for publication to the trade. The Associated Booksellers met the proposal with imaginative sympathy and appointed David Roy to serve with Walter Harrap in determining the conditions and acting as observers. The object was twofold: to determine, first, whether the British public would react favourably to coupon advertising and, secondly, to what extent it was detrimental to the interests of booksellers; and the books chosen were six volumes of a 'Living Languages' series to be published by Nicholson & Watson, who undertook to spend a minimum of £500 on coupon advertisements and during the effective life of these advertisements to refrain from any other form of advertising except showcards for use in bookshops.

[1] Founded by Walter Harrap; now The Publishers' Publicity Circle.

The report which the two observers submitted in April 1939 was published in summary as follows:

STATISTICS. The six volumes of 'Living Languages' were published in September 1938. The papers and magazines chosen were *John O'London's Weekly* (5 insertions), *The Listener* (5 insertions), *Strand Magazine* (3 insertions), and *World Wide Magazine* (3 insertions): three distinct types of copy were used. The total cost of advertising (excluding showcards and sales letters) was £520.

The value of coupon sales at full published price was £83. 6s. 6d. and trade sales £1,284. 16s. taken at an average discount of a third off (£1,927. 4s. published price). The percentage of coupon sales to total sales, therefore, was approximately 4%.[1] The orders on subscription (September 15th) amounted to 8,715 copies, of which number 3,333 were purchased by wholesalers, exporters, and the foreign trade, and repeats (up to December 29th) 2,814 copies. The percentage of coupon sales to total repeat sales was approximately 10·5%.

The subscription orders—averaging less than 1,500 per volume—showed that, as agreed, no unusual support was given to the scheme by the trade. The cost of advertising worked out at approximately 11d. per copy sold, or approximately 38% of the price received.

CONCLUSION. It is our opinion that the scheme was adversely affected by the political situation and that the books in question by nature of their subjects did not provide the perfect test. On the other hand, by including six books in the scheme and advertising them at the same time, the chances of success were increased.

The direct coupon response to the advertising was so small that booksellers should be satisfied that they have nothing to fear from similar schemes, if properly organized and if the advertisements are fairly worded so that no undue pressure is put upon the public to order direct from the advertisers.

This experiment in coupon advertising would seem to confirm the experience of subscription booksellers that the direct selling of books through press advertising is costly and cannot become profitable unless the advertiser has a large margin of gross profit to work upon.

It seems quite clear that this method of sales promotion cannot be applied in England with the same success as in America. America, with huge territories, inadequately covered by the retail bookseller, lends itself to the direct appeal, whereas in this country, so well served with bookshops, the public prefers to examine books before committing itself to a definite purchase.

Finally, in so far as the experiment can be taken as a criterion, we are of the opinion that, although the experience of Messrs. Simon & Schuster in America differs considerably from that of Messrs. Nicholson & Watson in this country, their contention that the bookseller profits rather than loses by the coupon type of advertising is fully justified.[2]

[1] 3% seems to be regarded in the 1960s as a good return for direct mailing, given the coincidental bookshop sales.
[2] *Members' Circular* XV, no. 4 (April 1939), 34-5.

It was unfortunate that the Munich crisis had coincided with the experiment and that the conditions which followed were not appropriate to further tests. But it may be noted that in America at least one publisher came to argue that the cost of his mail order department obliged him to offer a lower rate of royalty on sales made direct at the full published price than on sales to booksellers at trade price.

TWOPENNY LIBRARIES

The Joint Committee of 1927-8 had concluded that book borrowing was not generally conducive to book-buying and a new development in borrowing aroused bookselling scepticism again in the 1930s. The growth of small circulating libraries, the twopenny libraries as they came to be called, began in the summer of 1931 and by the end of 1932 the Joint Advisory Committee had listed as eligible 'other traders' sixteen firms, some of them—notably Messrs Salmon and Gluckstein—owning multiple shops. It must have seemed to some booksellers that their territory was being invaded by the tobacconists and at their Chester Conference in the summer of 1932 the Associated Booksellers viewed with considerable apprehension the rapid growth of small commercial libraries 'not connected with the book and allied trades'. Nevertheless the number grew, with 42 new entrants recognized in 1933, 77 in 1934, and 91 in 1935, and although a joint committee of publishers and booksellers was set up to study the effect of these libraries in 1936, it made no recommendations and new libraries continued to be recognized, although in diminishing numbers, until the outbreak of the war. They provided for some publishers a new impetus to 'partial remaindering'[1] and in June 1934 the Council of the Associated Booksellers asked the Council of the Publishers Association if it did not think that there was something wrong with a system which enabled one bookseller to offer 7s. 6d. novels to circulating and public libraries for 2s. 9d. while the rest of the trade were compelled to continue to maintain the full price for the same books. Required to give a ruling on a practice of which one of its own members was a persuasive advocate, the Council of the Association acknowledged that 'certain publishers by partially remaindering their stock of a book had, in effect, two trade prices for it, one enabling one bookseller to buy and sell at much lower rates than another', but sympathy without hope of a remedy was all that it could then offer to the Booksellers' Council. Practical action had to wait a little longer.[2]

[1] see p. 102.   [2] see pp. 199-200.

'THE JOINT COMMITTEE' REVIVED

If the two sides of the trade did not always see eye to eye on book clubs, Book Tokens and cigarette coupons, these new experiments in direct selling were at least discussed with greater frankness, and solutions of older recurring problems were easier to find. 'The Joint Committee' of 1927–8 was revived in a single session[1] in October 1932 when the interpretation of the Net Book Agreement, the recommendations on remainders and on allowances to booksellers upon the publication of cheap editions, the definition of 'second-hand', and the sale and marking of review copies were among the questions reviewed. The session closed with the following six-point resolution:

(1) That in the present times of economic depression the book-buying public could not afford large and expensive books and therefore attention should be directed to books of lower price.

(2) That it might be possible, by means of greater co-operation, for booksellers to shoulder some of the present heavy burden of publicity on publishers, and in particular that the Booksellers Association should urge its members to take steps to ensure a more systematic and effective use of publicity material provided free by publishers.

(3) That too much publicity was directed towards the pushing of new books and that it might be advantageous if some of the older and well established books received more attention.

(4) That there should be an increase in the practice which some publishers have adopted of sending early copies of books to booksellers so that they could be able to place their orders more exactly and to advise customers.

(5) That while in no way challenging or altering the publishers' admitted rights to find their own channels of distribution it would be an advantage if publishers would, where practicable, consult with booksellers before taking up any new methods of marketing, and publishers are reminded of the importance of distinguishing between new business and the mere divergence of existing business into new channels.

(6) That at present attention to increase of turnover is more important than improvement of terms.

Less formally than in this revival of 'The Joint Committee' Stanley Unwin and Basil Blackwell during their contemporary Presidencies not only began to hold regular fortnightly meetings, but also in September 1934 acted as joint hosts at an unofficial week-end conference of some fifty publishers and booksellers at Ripon Hall near Oxford.

The interpretations of the Net Book Agreement were tightened in 1937

[1] The P.A.s representatives were Bertram Christian, W. Longman, Stanley Unwin, Sir F. Macmillan, Jonathan Cape, Percy Hodder-Williams, W. G. Taylor.

by the addition of two clauses designed to check breaches of the Agreement by certain library suppliers and by dealers in review copies who were accustomed to sell, as if shop-soiled or second-hand, new net books at small discounts. The marking of review copies had been a subject which had aroused strong differences of opinion since in 1907 the Association had recommended all its members to stamp all review copies indelibly with the words 'Presentation copy'. In 1935 the Council advocated the use of a machine to perforate pages with the letter R, and again in 1937 the activities of certain reviewing agencies, whose interest was in sale rather than review, and of a bookstore in the east end of London, which showed remarkable ingenuity in obtaining underground supplies of net books for sale at discounts, led the Council to try to secure general agreement to mark unobtrusively all review copies except those in specialist subjects. The object was twofold: to discourage public and circulating libraries from buying review copies, and to stop the sale as review copies of books which were nothing of the sort. But dislike of defacing a book and fear of antagonizing reviewers made a few important firms unwilling to commit themselves and an agreement had not been achieved when war broke out.

THE JOINT ADVISORY COMMITTEE AT WORK: 'OTHER TRADERS', BOOK AGENTS, QUANTITY DISCOUNT SCHEME

The most important outcome of 'The Joint Committee' had been, as we have seen,[1] the formation in 1929 of a standing Joint Advisory Committee to be responsible to the Publishers Association for the 'recognition' of booksellers and the observance of the Net Book Agreement. When in the following year the President of the Board of Trade and the Lord Chancellor set up a Committee on Restraint of Trade, the memorandum which the Association was invited to submit upon its functions and methods of working (and which the Committee accepted without question) was concerned particularly with the work of the J.A.C.[2] That the J.A.C. was able quickly to get through so great a volume of work as was involved in reviewing and classifying all existing exporters and that its authority came quickly to be accepted was due in no small measure to its first two chairmen, William Longman and W. G. Taylor. Of its main work, the consideration of applications for eligibility for supply at trade terms, the year 1937 may be quoted as an example: 300 applications were investigated and 140 were rejected. 'It is' wrote Wren Howard in the Annual Report on

[1] see p. 102.
[2] The memorandum is printed in *Members' Circular* VIII, no. 16 (May 1930).

that year, 'somewhat difficult to believe that there is room for as many as 160 new booksellers and [commercial] librarians in the trade as we know it today, and the fact that so many applications were granted indicates that no application representing a genuine attempt to further the sale or distribution of books is discouraged.' As an alternative to full recognition an applicant might be accepted as an 'other trader' eligible for trade terms on a limited class of book (e.g. a sports dealer for books on games). In 1933 applicants apparently intending to trade only in a very small way began also to be listed for supply, not by publishers direct, but by wholesalers only; and the same year saw the introduction of the book agents scheme, under which an applicant with a limited interest in the sale of books (e.g. a bookstore in a technical institution or a bookstall in a church) or an applicant with a temporary interest might be licensed to be supplied at a discount by named booksellers.[1] In 1937 a quantity book-buying scheme also was put into operation, under which a person or body wishing to give away a 'large' quantity of a specific net book for philanthropic or propagandist purposes or in connection with their business could be supplied by an authorized bookseller at a discount; and at that time with its different values 500 copies of a 6*d*. book or twenty-five copies of a 10*s*. 6*d*. book were given as examples of 'large' quantities.

The consideration of applications from Public Libraries for licences under the agreement with the Library Association also fell on the J.A.C., except between the years 1934-7 when a separate Library J.A.C. was in existence.[2] The first applications were considered in October 1929 and at the end of 1934 out of a probable total of 377 eligible libraries 346 were holding licences. It was not until 1945 that it could be claimed that no library of any note remained outside the scheme.

THE NET BOOK AGREEMENT AND THE 'CO-OPS'

In these ways the Association, acting in concert with the Associated Booksellers through the J.A.C., sought to maintain without irrational restrictions the principle of net prices. Infringements of the Net Book Agreement were generally inadvertent or of minor significance, but in 1934 a serious breach by a Birmingham firm led to the withdrawal from them of trade terms on net-priced books for more than seven months and the persistent and ingenious defiance of the Agreement by a bookstore in

---

[1] The book agents scheme was, in effect, first applied in the agreement with the Library Association (see pp. 96, 104-5). The wider use was much advocated by R. F. West and Basil Blackwell.

[2] It had the benefit of William Longman's chairmanship throughout its existence.

London, which I have already mentioned, caused the listing of its many aliases during a period which began in July 1936 and—such were the underground sources of supply which it was able to tap—lasted throughout the war.

The problem of the sales of books by the Co-operative Societies and the allowance of a dividend on sales of net books, which had remained beneath the surface since 1927,[1] also re-emerged in 1934 when the Gloucester branch of a large multiple store threatened to meet the competition of the local Co-operative Society by itself allowing a dividend on books. A joint committee of publishers and booksellers which was then set up recommended that all the Societies, both wholesale and retail, should be treated as ineligible for trade terms until they had signed the Net Book Agreement and had given an undertaking to treat net books as proprietary articles and exclude them in the calculation of their dividends. An amicable solution could not be reached with representatives of the Wholesale Society and eventually in April 1939 on the recommendation of the Juvenile Group, supported by other publishers of children's annuals outside the Association, the Council decided to stop the supply of net books at trade discounts to all Co-operative Societies. Within twelve months 150 Societies had signed the Net Book Agreement and, although the Wholesale Society remained in opposition, it was clear that the retail Societies generally accepted the need to maintain book prices.

BOOK WEEKS AND EXHIBITIONS

If in its administration of the Book Tokens scheme the National Book Council found itself subjected to opposing pressures from publishers and booksellers, it provided common ground in its organization of book weeks and exhibitions. In the winter of 1934–5 successful exhibitions were held in Liverpool, Manchester and Glasgow and in the following years the organization, in conjunction with local booksellers, of book weeks and permanent exhibitions in Public Libraries became a feature of the N.B.C.'s work in the provinces. In London also it undertook the organization of the succession of exhibitions sponsored by the *Sunday Times*, which from a small beginning at Sunderland House in 1933 grew to a national book fair at Earls Court in 1938. The N.B.C. was also responsible for the production in 1936 of two documentary films about books, *Cover to Cover* and *Chapter and Verse*.

[1] see p. 97.

RE-ORGANIZATION AND NEW RULES

In these years the Association had to give thought to its own organization and to the insufficiency of its secretariat to deal with the work demanded of it. At the Annual General Meeting in 1931 G. H. Bickers called attention to the need for re-organization and at the following Council meeting a committee of investigation was appointed.[1] The committee's report, which covered the secretarial work for the Groups and the Book Trade Employers' Federation, which was still being done by Harvey Greenham in his chambers in King's Bench Walk, and the need for a single larger secretariat and more space to house it under one roof, was approved at a special General Meeting in August 1932;[2] and before the end of the year more accommodation had been secured at Stationers' Hall and F. D. Sanders had been appointed Assistant Secretary. It was an appointment of the greatest importance for the future of the Association; that the right man had been found was quickly recognized in his appointment as Secretary from 1 January 1934. There is now no President alive to remember how the Association existed before Frank Sanders's arrival. He was to be a tower of strength to the trade and the friend of many publishers. William Poulten, who had been Secretary from the beginning of the Association almost forty years before, continued for a period as Consulting Secretary in name.

The Association also amended its Rules in 1935 to provide that no member of the Council who had served for six consecutive years, excluding any period of service as an officer, should be eligible for re-election during the year immediately following. But the constitution needed more revision to make the Association a corporate body and so to enable it to act more quickly and sometimes less cautiously in defence of the Net Book Agreement and to protect the officers and individual members of the Council, Groups and Committees from personal liability for acts of the Association. A revising committee[3] was set up in September 1935 and although it had the benefit of the experience of the Associated Booksellers, who had almost completed a new constitution of their own, its task in meeting the

---

[1] The Officers (Bertram Christian, W. Longman, Stanley Unwin), Harold Macmillan, Scheurmier, H. G. Wood (Nisbet's).

[2] The annual subscription of £8. 8s. to the Association and a graded subscription, according to number of employees, to the Federation were maintained unchanged; the additional subscription to the Educational Group was reduced and the subscriptions to the other Groups were abolished.

[3] The Officers (W. G. Taylor, Stanley Unwin, G. Wren Howard), J. H. Blackwood, Scheurmier (with power to add).

demands of members and of the Registrar of Friendly Societies was more formidable than at first appeared and the new constitution was not in force until 1939.

THE GROUPS, NEW AND OLD

The adoption of the report on re-organization at the General Meeting in August 1932 brought into being a new Group. Since 1925 the Book Trade Employers' Federation, with Harvey Greenham as Secretary, had been the body responsible for negotiating wages and conditions of employment with the appropriate union and since the General Strike in 1926 it had had no serious problems with which to contend, for the Union had withdrawn from the trade. Now it was agreed that the Federation should be merged with the Association, which would accept its obligations to the union and act in future through an Employment Group. Harold Macmillan, who had been the last chairman of the Federation, became the first chairman of the Group.

These years also saw the beginning of the Indian Group, as we have seen,[1] and of the Map and Medical Groups. In 1934 the Director General of the Ordnance Survey put forward a recurrent claim that all maps, no matter how small their scale, were subject to Crown copyright if Ordnance Survey material had been used in their reproduction. Fears that this claim might be applied restrictively and that the Ordnance Survey might itself compete with the private firms in the production of small-scale maps for the educational and commercial markets were intensified by the publication in August 1936 of new regulations governing the reproduction of O.S. maps, and in March 1937 a conference of map publishers requested the Council of the Association to authorize the formation of a Map Publishers Group. It met for the first time in November, with George Philip in the chair, and was satisfied that its fears were unlikely to be realized. The Medical Group held its first meeting in April 1938, with R. F. West as its first chairman, and was soon active against a threat of Government publishing within its field.

Of the older Groups, the stand of the Juvenile Group against the Goliath of the 'Co-ops' has already been recorded. Less laudably, in 1938 the Group was pressing for action against the importation of cheap Japanese books and it revealed that some firms, often not in membership of the Association, were marking newly published books with fictitious prices, i.e. prices far above those at which the books were intended to be sold. The Educational Group continued to meet with the greatest

[1] see p. 122.

regularity and of its activities the following deserve record: the first recorded meeting with officials of the Board of Education in 1932 to discuss the teaching of particular subjects and the listing in its official publications of books useful to teachers; the dissuasion of three authorities, Newcastle upon Tyne, Warwickshire and the West Riding, recognized for supply at trade terms, from selling to pupils in maintained secondary schools and to county libraries; and corporate action in 1938 to deny the use of copyright material in school books published by the state Governments in Australia or to ensure, by the charging of heavy copyright fees, that state-publishing would not be able to compete at advantageous prices. Both the Educational and the Fiction Groups were occupied in 1937-8 with problems arising from increased manufacturing costs and both secured co-operative agreements among their members relating to increases of published prices. The Fiction Group also reached an understanding with the four principal Circulating Libraries[1]—Boots, Harrods, W. H. Smith's, and *The Times* Book Club—that only novels of less than 90,000 words would continue to be priced at 7s. 6d. and that they would accept novels of 90,000 to 110,000 words at 8s. and of more than 110,000 words at 8s. 6d. In December 1938 the Group submitted to the Council a memorandum on partial remaindering and in the following March a committee was appointed with Sir Humphrey Milford as chairman to investigate the practice; but war conditions necessitated the abandonment of the inquiry.

THE BOOK MANUFACTURERS' ASSOCIATION

It was not only the Educational and Fiction Groups which were concerned with increasing costs of manufacture. It would probably have been conceded generally that bookbinding was an insufficiently profitable industry, but in their attempt to raise the level of prices the leaders of that industry went about it in a misguided way. Early in 1934 publishers became aware of an organization called the Bookbinders' Costs Investigation Committee and in March the Council of the Association addressed a letter to the secretary of the Committee asking whether it was true that an association of binders and clothmakers was in course of formation with the object of controlling binding prices by coercive means and whether representatives of the Committee would be willing to answer questions. The answer came that the Committee had been converted into The Book Manufacturers' Association and that any safeguards which the Publishers Association wished to suggest could be discussed at a meeting. In its turn the P.A. sought a denial from the B.M.A. that it was inviting the cloth

[1] Mudie's famous Library came to an end in 1937 after ninety-five years' existence.

manufacturers to exert pressure on 'recalcitrant binders' by cutting off their supplies and stated that a meeting would be advantageous only if the cloth manufacturers had withdrawn from the B.M.A. Throughout the ensuing negotiations, which continued with some acrimony until the summer of 1935, the P.A. was not given reason to change its belief that the Winterbottom Book Cloth Co., which held what was not far short of a monopoly, was the policeman of the B.M.A. and as the months went by evidence grew that Winterbottom's small and enterprising rival, Thomas Goodall & Co., which had stood out of the B.M.A., was doing less business. At a meeting in October 1934 with master bookbinders, some members and some not members of the B.M.A., the representatives of the P.A. believed that constructive progress had been made and that, although the following resolution had not been drafted during the meeting, it represented its intention:

> that a joint Committee composed of three publishers and three master binders, two of the latter being signatory members of the Book Manufacturers' Association and the other being a binder not in membership of the Book Manufacturers' Association, be set up to examine fully and *de novo*, and report to the bodies concerned, the position of the binding trade, its reported uneconomic condition, the nature of remedies which might be adopted to the benefit of the trade by such a joint Committee of publishers and binders, and in particular the steps that might be taken to eliminate ruthless price-cutting below actual cost.

But the representatives of the B.M.A. disputed the accuracy of the resolution as an expression of agreement reached at the meeting and any hope of the joint Committee was deferred while the B.M.A. instructed a firm of accountants to investigate binding costs and to report on a proposed schedule of prices.

On 13 July 1935 *The Times* printed a prominent report upon the initiation by the B.M.A. of a policy for the maintenance of fair prices for bookbinding. A schedule of minimum prices, it said, had been worked out and had been scrutinized by an independent firm of accountants as an assurance to customers that the prices were the lowest that the trade could charge if a reasonable profit was to be made. 'For a long time' wrote *The Times*, 'book manufacturers have realized that prices were too low. Machinery was not fully employed; competition for orders exerted a continual downward pressure, and reductions of price had been made even when profits were non-existent. The alternative to this destructive competition was a scheme of minimum prices.' And *The Times* concluded its report with the following prognostication: 'There was at one time a proposal that pressure should be put on firms not adhering to the scheme through the control of raw materials, but it is now thought that with the co-operation of publishers, such a step will not be necessary.' To this announcement the P.A. made a rejoinder which was inevitably rather feeble if continuing discussion between the two Associations was not to be prejudiced and it was given an insignificant position in *The Times*. The B.M.A.'s proposals, it said, were subject to discussion between bookbinders, by no means all of whom were members of the B.M.A., and the P.A. and it could only be said that a majority of book-publishers did not agree with *The Times*. In fact, without knowing it, the P.A. had already won. No further discussion was held, the B.M.A. ceased to exist, at any rate so far as the P.A. was concerned, and bookbinders continued to live an unprotected and precarious existence.

As throughout 1938 and 1939 the inevitability of war increased, the sale of books, as of much else, diminished; and, as will be seen in the next chapter, the choice of these two years as reference years was to accentuate the severity of paper rationing as they became less and less representative. Although the 1930s ended, as they began, in a period of poor trade, it was a decade of inventiveness. These were the years not only of the invention of Book Tokens, book clubs and book fairs, but of the first use, by Victor Gollancz, of the Sunday papers for bold advertising of his best-sellers and of Allen Lane's first hatching of Penguins, to become the progenitors of many other paperbacks of varying kinds and sizes.

9
1939-1946 (1)
World War II: the book front

That the peace bought at Munich was a precarious one the Association was not long in doubt. Early in 1939 the Council was considering the Ministry of Transport's proposals for the pooling of commercial vehicles in the event of war and was trying unsuccessfully to persuade the Ministry to favour pooling by groups within a trade rather than within a local authority area. In March Geoffrey Faber, taking office as President, looked forward to the year ahead as one of cautious consolidation, but nevertheless proposed the formation of a War Emergency Committee. By the early summer the Committee was at work, formulating recommendations for the guidance of members in their treatment of employees called up for service in the Territorial Army and the Militia and was considering the question of evacuation. Since the main distributing book-centres (W. H. Smith's and Simpkin's) were remaining in London, it was evident that no general plan for the evacuation of publishing offices was practicable, even if any suitable and unappropriated centre for the trade could have been found. On 13 July the Civil Defence Act came into force, requiring employers to strengthen basements or otherwise provide splinter- and blast-proof shelters, and to train and equip teams for the Air Raid Precautions service. Steel helmets were purchasable for 8s. 6d. By the closing days of August, principals and their staffs were filling sandbags, strengthening window-panes, and improvising 'black-out' precautions.

'THE PHONEY WAR'

After the precarious peace came the seven months of 'phoney' war, falsifying all expectations. Book sales, which had been running at a low level since the Munich crisis, were further depressed in September, but had recovered by Christmas in spite of—or probably because of—the black-out; and indeed until the end of the war publishers' difficulties were to be of supply and never again of demand. With one exception, the trade was hardly aware during these months of the sacrifices and restrictions which it was to be required soon to face.

War risks insurance. The exception was occasioned by the War Risks Insurance Act, 1939. The Government's decision that all traders carrying stocks of saleable goods valued at £1,000 or more must insure against war risks placed a heavy burden upon publishers and booksellers and presented the trade with its greatest co-operative problem since the establishment of the Net Book Agreement. The premium of $1\frac{1}{2}\%$ for the initial period of three months fell with great severity upon the book trade because of the slowness with which book stocks are turned over. The Board of Trade had the power to rule that a particular class of goods was non-insurable under the Act and, if the trade had been able to agree upon a joint application by publishers and booksellers that books should be ruled non-insurable goods, it is to be presumed that the Board of Trade would have consented to make such a ruling. In that event, unless the trade had formed a pool of its own, as some members indeed advocated, those firms such as Longmans and Allen & Unwin which were to lose millions of books in the air-raids of 1940-1 would have been in dire financial difficulties. As it was, at a special General Meeting on 18 September 1939 a motion requesting the Board of Trade to declare books to be uninsurable while in the ownership of the publisher was carried, but only by 46 votes to 39 in a ballot after a show of hands had rejected it by 39 to 38; and on the following day the Associated Booksellers in a special General Meeting decided that they must be covered by the scheme. In such a confused situation the Board of Trade could not have made a ruling, nor was it willing to exercise the power which it appeared to have, of varying the rate of premium to meet the different circumstances of different trades. A scheme for recovering the cost of the premiums by an adjustable W.R.I. surcharge on published prices was then proposed by the Council and first rejected at a General Meeting on 9 October, then approved at another on 30 October and worked out in laborious detail with representatives of the Associated Booksellers, and finally rejected again on 17 November in consequence of a change in the attitude of the Booksellers. Just as booksellers were divided as to the wisdom of immediately passing on the premiums to the book-buying public, so was there a division between the general publishers, who feared the effect of a surcharge, particularly on Christmas sales, and the educational publishers, who could pass it on with less apprehension. An alternative attempt to meet the difficulties of the booksellers by the grant of a special W.R.I. discount failed also to obtain the general support of publishers, many of whom felt that they could not undertake to bear a double burden. The Association had no alternative but to leave its members to their own devices.

The subsequent history of the War Risks Insurance scheme, so far as the Association was concerned, was not extensive and it will be convenient here to continue it. The operation of the scheme by individual publishers continued to bristle with complexities. Month by month each publisher had to strike an insured figure for continually moving stocks of every title in his list, with varying values for new books and old books and with the whole subject to the 'average' clause. The rate of premium which had started at 6% p.a. soon dropped to 3% and, when bombing started, rose to $4\frac{1}{2}$% of the insured value. If that value were based, as it normally was, upon balance sheet values, the compensation obtained, especially in respect of old stock, was liable to prove miserably inadequate. Even if a publisher insured the stock of a new book at cost and then lost, say, 100 copies the Government assessor might argue that a reprint of, say, 2,000 copies would be normal and that the cost of replacing the 100 copies would be one-twentieth of the reprint cost. Furthermore, if the compensation obtained exceeded the balance sheet value of the stock destroyed, the excess (unless employed in replacing the stock within the company's year, which was frequently impracticable) figured as profit and was liable to income-tax and the excess profits tax. In February 1941 the Council proposed to the Board of Trade that a lower rate of premium should be payable on older stock and that the Board of Inland Revenue should be urged to rule that compensation paid on books published in a year previous to that in which the loss occurred should be treated not as though it were a sale, but as a return of capital or at least be spread over several years. But the Board refused and the anomaly continued. As it became harder and harder to find the paper and a printer and a binder to replace stock quickly and as publishers became increasingly liable to excess profits tax at 100%, their W.R.I. premiums seemed to be a form of taxation rather than insurance.

Paper. Although there were to be moments in the war when the maintenance of supplies of books was threatened not by scarcity of paper, but by the withdrawal of men from printing and women from binding, and even of girls from the cloth mills, paper was to be the most constant anxiety and it was particularly to the anticipated control of paper that the President, Geoffrey Faber (supported by the Vice-President, G. Wren Howard, with his exceptional knowledge of book-production) and the Council applied themselves during the first seven delusive months of the war. At the beginning of August 1939, in consequence of a secret memorandum on the measures which the Ministry of Supply was about to take for the control of paper, publishers were asked to furnish the Secretary of the Association in confidence with particulars relating to their consumption of paper; and the majority of publishers complied. The organization of the Paper Control as first devised by the Ministry of Supply, while giving representation on various committees to other consumers of paper, ignored the existence of book-publishers. The President and Vice-President in an interview at the Ministry in September succeeded in obtaining three representatives on Advisory Committee No. 3 (General Printing) and one on No. 5 (Boards).[1] But in November the publisher representatives were protesting that they alone of the members of Committee No. 3 had not been consulted by the Paper Controller before he had agreed to an increase of 40% in the prices of paper.

In an attempt to achieve either the exclusion of book-publishers from the rationing of paper and boards or the automatic release to them of supplementary supplies, the Association obtained in December a remarkable response to a proposal for voluntary rationing by the reduction by every publisher of his consumption of paper and boards to 80% of what it would have been in peace time, given a similar output; and in February 1940 the Association, trying to act for a trade which had consistently rejected any peace-time demand for statistics as an unjustifiable inquisition, obtained its first rudimentary figures of book exports. They showed that scarcely any publishing house derived less than 15% of its sales from exports, that the large firms with extensive and long-established connections overseas exported between 50 and 60%, and that the average might be $33\frac{1}{3}$%. The ignorance and the surprise with which these high figures were received reinforced the need for more, and for accurate, mass statistics and not just a bundle of incommensurable percentages. In spite of these two strong arguments books were not excluded from rationing when it came into force on 3 March 1940, but the ration was 60% of consumption in the twelve months to 31 August 1939, which was to be the immutable reference period throughout the war. (The total consumption was given as 45,000 tons,[2] which had grown to 52,000 tons by the end of the war, presumably because some users did not at first establish their right.) Books were also listed by the Controller as one of

---

[1] The representatives were: on No. 3, G. O. Anderson, Richard de la Mare (Faber's), F. J. Martin Dent (Dent's); on No. 5, G. Wren Howard.

[2] A ton of paper would have produced about 3,000 copies of a novel of 250 pages.

the essential purposes for which additional paper could be licensed; and at the February Council meeting the President expressed some confidence that additional licences up to 20% would not be difficult to obtain. But the invasion of Norway and the cessation of imports from Scandinavia were still to come.

Labour and other restrictions. During this first period of the war the Association was active also in trying to persuade the Ministry of Labour to include in the Schedule of Reserved Occupations (which had been originally issued early in 1939 to indicate persons who were not free to join voluntary service organizations and had become the basis for calling up for military service) occupations essential to publishing; and it secured an assurance in March 1940 that seventeen listed occupations were applicable to the book trade, and the subsequent addition of three special categories: editorial staffs and production clerks (book-publishing), both with reserved age thirty-five; sales representatives (book-publishing), reserved age thirty.

In general Government Departments showed themselves anxious to enlist the help of the Association in order to make restrictions as little burdensome as possible. In order that funds and shipping should be available for the purchase abroad and the transport of munitions and raw materials, the import of books, as of other manufactured goods, became subject to licensing in September 1939; and it was largely to the credit of the Association's Treasurer, Walter Harrap, that the scheme adopted by the Import Licensing Department ensured that applications were intelligently and quickly dealt with. A system of export licensing was set up for a different reason: to prevent the use of exports by enemy agents as a means of conveying military intelligence out of the country; the method of using general licences to approved firms was adopted from the outset. Petrol rationing brought no serious inconvenience so far as trade deliveries were concerned, but the Petroleum Department was unsympathetic to the needs of travellers and educational representatives.

Exports acquired additional importance and some new impediments. New Zealand was compelled by balance of payments difficulties to reduce imports of books by 50% and representations to Walter Nash, the Finance Minister, were unavailing. An inevitable increase in the price of colonial editions of novels brought reduced orders from Australia; and shipping delays to Canada in September and November 1939, with others to come, were to have serious effects on British publishers' fortunes in that market.

Faber's conclusions. In the Annual Report of the Council in March 1940 the President, Geoffrey Faber, drew certain conclusions from the Association's activities in the first seven months of war. The first was that the war had greatly strengthened the importance of the Association and its usefulness to its members, on whose behalf the Officers had been able to present to Government Departments considerations of vital concern to every publisher solely by virtue of their status as the official representatives of the Association. The second was that book-publishers of repute could not afford to remain outside the Association and that it was unreasonable that they should participate in the advantages, but not in the obligations. Yet the question also arose how the Association could accept responsibility for members whose business was only in part concerned with the publication of books—a question to which one answer, as will be seen, was to come from outside in the following year. A third, and more controversial, conclusion was that the old individualism traditional to publishing needed some modification, and Faber instanced the absence of trade statistics, which so greatly handicapped the effective presentation of argument to Government Departments, and the recent inability of the Association to agree upon a common policy about war risks insurance. 'These observations' Faber wrote, 'do not mean that the Council is preparing to execute a *coup d'état* or to set up a benevolent dictatorship. They do, however, mean that in the opinion of many publishers the time is coming when the will of a sufficiently large majority about matters of common importance should be binding upon all.' The individualists, mistakenly raising the banner of 'no censorship', were to fight a prolonged skirmish on the field of paper; and a hint of benevolent dictatorship had recently been scented.

The Association's set-up. In April 1939 a War Emergency Committee had been appointed. When at the beginning of the war it appeared probable that frequent consultations on urgent matters would be necessary, the Committee was somewhat enlarged; and it held several meetings. It was soon evident that unless it was given special powers it could not relieve the Council of the responsibility for retraversing the same ground, and at the General Meeting on 9 October it was agreed that due notice should be given of a new regulation defining the powers of the Committee. But other advice prevailed and in the event it continued to be possible to call special meetings of the Council itself on necessary occasions at short notice. A special Emergency Committee remained in being without special powers.

It may be convenient here to summarize the organization with which the

'*The phoney war*' 165

Association armed itself at the beginning of the war proper. In addition to the special Emergency Committee,[1] a Public Relations Committee,[2] an Economic Relations Committee,[3] and a Paper Committee[4] were set up, ceasing gradually to exist as their work was done or as it passed to other committees which were formed for specific tasks. Geoffrey Faber's Presidency continued until March 1941; he was followed by Walter Harrap, by R. J. L. Kingsford in 1943 and by B. W. Fagan[5] in 1945. But the work of these Presidents would have gone for nothing without the outstanding ability, the tact, the cheerfulness, and the apparently tireless energy of the Secretary, Frank Sanders, and the support which he had from his staff. Of the many civil servants with whom these Officers were to be in constant contact the intellectual power and the integrity of two in particular survive in one memory: H. J. (now Sir Herbert) Hutchinson, Permanent Under Secretary at the Ministry of Supply, and Richard Pares, temporary Assistant Secretary at the Board of Trade.[6] Less favourable recollections of one Minister remain.

1940–1941

Within a few weeks of the beginning of paper rationing in March 1940 the enemy's invasion of Norway cut off imports from Scandinavia and within a few weeks more the capitulation of France ended the supplies from French North Africa of esparto grass, from which 65% of the papers used by British book-publishers was made. The maintenance of the paper quota, the recognition given to books by the public and by Parliament in their eleventh-hour exemption from the New Purchase Tax, and the destruction of Paternoster Row and its precincts by bombing on the last Sunday of 1940 were to be the principal events of the year.

Paper. In April a Paper Control Order reduced the ration from 60% to 30% (13,500 tons p.a.) and revoked outstanding licences by 50% and the Controller gave warning that the ration might be not more than 15–20% before the end of the year. Strong representations made in memoranda to

---

[1] The Officers, R. P. Hodder-Williams (Hodder & Stoughton), W. G. Taylor, Stanley Unwin.
[2] G. Bickers, A. S. Frere (Heinemann), R. J. L. Kingsford, Harold Macmillan, Sir Humphrey Milford, Gerald Rivington, R. F. West.
[3] Messrs Jonathan Cape, Collins, Dent, Hodder & Stoughton, Harrap, Macmillan, Oxford University Press; these member-firms to send to each meeting the representative best qualified for the business to be discussed.
[4] The Officers, W. A. R. Collins, R. de la Mare, Daniel Macmillan.
[5] of Edward Arnold & Co.
[6] Fellow of All Souls College, Oxford; subsequently Professor of History at Edinburgh University.

the Export Council and the Paper Control and in a letter from the President to Herbert Morrison, the new Minister of Supply, were successful and on 2 June the Minister replied that although there was not enough paper for the requirements of the services and for important civilian needs, nevertheless the quota for books was to be restored to 60%, and 'I hope' he said, 'that your Association will feel that, in the circumstances this is evidence of the importance which is attached to the maintenance of the book trade'. There was also the possibility of additional licences for educational books, for which certificates would be required from an Educational Advisory Panel[1] to be set up by the Association.

The Export Group. The need for exports to help to pay for the war led the Government to establish an Export Council and the preparation of an export drive made it necessary to explain to that Council the peculiar circumstances of the book-publishing trade. This was done in a memorandum of 5 April 1940 which demonstrated the close dependence of the export trade in books upon the home trade and the impracticability of manufacturing 'lines' of books for export. Its main contentions were accepted and it was agreed that no grandiose export drive for books was possible. It was nevertheless considered necessary that an Export Group for the book-publishing trade, as for other trades, should be organized on the basis of the Association. The Group was officially recognized in May 1940, the Officers and Secretary of the Association filling the corresponding offices of the Group. Within the year the membership rose to 369, as compared with 142 members of the Association, and one General Meeting and two Executive Committee meetings were held. The Group handled a wide variety of matters affecting exports, such as timber for packing cases, permits for overseas travel, and exchange regulations;[2] and the Export Council[3] was of no less value in supporting the claims of the interdependent home trade for reasonable protection. But the principal significance of the Group was, first, that the 60% paper quota, with the possibility of additional licences to firms with an unusually high proportion of exports, became obtainable only by its members, non-members being entitled to 40%; and, secondly, that its members were required by the Board of Trade to make quarterly returns of their total and export sales

[1] W. G. Taylor (Chairman), the chairman of the Educational Group (H. G. Wood), Bertram Christian, R. J. L. Kingsford, R. F. West.
[2] A useful concession was that from July 1940 Form C.D.3 (payment for exports) was not required for books sent by book post.
[3] The executive member responsible for books was F. D'Arcy Cooper (created baronet 1941), chairman of Unilever Ltd.

figures. Thus the Association began to get trade statistics, and since the Export Group included all publishers of any consequence, the figures were comprehensive.

The rationing of paper at 60% and of boards at 50% continued until the end of January 1941 when the Association was notified that on the recommendation of a committee on raw materials under the chairmanship of Lord Burghley[1] a reduction to 50% and 40% respectively would be made in March; and the reduction could not be averted. The control of paper also began in mid-1940 to limit the distribution of catalogues and prospectuses. It was announced also that the consumption of cotton, which had already been reduced by 25%, was to be reduced by 75% from October, but the intervention of the Export Council secured the maintenance of adequate supplies of bookcloth.

The Blitz. The blackest point of the year came on 29 December 1940, when the rain of incendiary bombs on the City devastated Paternoster Row and Ave Maria Lane, destroying totally or partially the buildings and the stock of some twenty publishing houses, Stationers' Hall and the adjoining offices, several bookshops, and the wholesalers, Simpkin Marshall. So far as the Association was concerned, the old premises in Stationers' Hall Court had become inadequate and the move to 28–30 Little Russell Street, W.C.1, precipitated by the raid, had been foreseen. There the Association was joined by the Associated Booksellers, who had also been bombed out of Paternoster Row.

Simpkin's. The total destruction of Simpkin's stock brought into action the Economic Relations Committee to save this vital wholesale and export business from disappearance. The Committee had been set up to watch the effect of war conditions upon bookselling and under the vigorous lead of its chairman, Walter Harrap, it acted with great speed. As soon as the owners made it clear that they would put the business into liquidation, three members of the Committee, W. B. Cannon (Oxford University Press), Harrap, and G. Wren Howard, with Stanley Unwin, on their private responsibility made an offer for the goodwill. The offer was accepted a week after the fire. Simultaneously the Committee was in touch with I. J. (later Sir James) Pitman, of Sir Isaac Pitman & Sons. In the event, Pitman's generously took over the agreement to purchase and the entire financial and administrative responsibility for operating the new enterprise. The new Simpkin Marshall (1941) Ltd was a non-profit-

[1] now the Marquess of Exeter.

making co-operative company, designed to operate the warehousing and distribution of books at the lowest possible cost in terms of commission. The Association had the right to nominate four Directors on the board—and duly nominated the four who had made the original offer—and the shares were held by Pitman's subject to the right of the Association at any time to take them over on repayment only of the amount advanced by Pitman's. The scheme was generally welcomed by members of the Association in a Special General Meeting on 4 February 1941. But to this new Simpkin's, which was a distributing agency rather than a wholesaler holding stock and capable of placing large orders, some publishers remained opposed and some gave grudging support; and it faced an impossibly uphill start. The book shortage gave it no help; the supply of books to customers' special orders—a prime attraction of a wholesaler to a retailer—was generally impossible and famine conditions drove booksellers to seek favours from publishers direct. These conditions were to last until the rationing of paper came to an end in 1949 and the part which Simpkin's might then have played in the coming problems of distribution was not foreseen.

The purchase tax battle. If 29 December 1940 brought the blackest night, 13 August 1940 was undoubtedly the brightest day. It brought the first recognition by the Government that books were something more than a commodity of trade. In his April Budget speech the Chancellor of the Exchequer, Sir John (later Viscount) Simon, announced the Government's intention to institute a purchase tax; and the fight for the exemption of books began. Representations made to the Chancellor resulted in an interview on 10 May in which the President and Vice-President, Faber and Wren Howard, accompanied by Harold Macmillan and David Roy, of W. H. Smith & Sons, met Sir Wilfrid Eady and other members of the Board of Customs and Excise. After this not unsympathetic interview a special memorandum putting the economic case against the taxation of books was prepared and submitted to the Treasury and arrangements were made for a deputation of eminent men, led by the Archbishop of Canterbury (Dr Cosmo Lang), to put the cultural case directly to Sir Kingsley Wood, who by then had replaced Sir John Simon as Chancellor. The deputation, accompanied by the President and Vice-President, Lord Hambleden and Stanley Unwin, met the Chancellor on 21 June, but appeared to make no headway against his explicit refusal to differentiate between books and boots.

The one remaining hope lay in the mobilization of public and parliamentary opinion. Kenneth Lindsay, M.P., took charge of the campaign in the House of Commons and with Harry Strauss, M.P., tabled an amendment to the Bill excluding books from the scope of the tax. Meanwhile a breakneck public campaign was organized. On 5 July a well-attended public meeting was held in Stationers' Hall, with Sir Hugh Walpole in the chair, and a National Committee for opposing the taxation of books was set up. On 6 August a meeting of Members of Parliament large enough to fill the biggest committee room at Westminster was addressed by the Archbishop of Canterbury, J. B. Priestley (who had also spoken at the Stationers' Hall meeting and had taken an active part in the deputation to the Chancellor) and the President. It was clear by this time that Mr Lindsay's amendment would receive very weighty support in the House. But the amendment was never debated, for on 13 August the Chancellor announced that he had decided to exclude books from the Bill. Their exclusion established the case that 'books are different'—a precedent which was to be of great value in many subsequent representations to authority.

It would be impossible to name all those who shared in the credit for this victory. Besides those already mentioned, the Council put on record that Sir Arthur Eddington, O.M., Professor A. V. Hill, Dr Albert Mansbridge, Sir Charles Grant Robertson and Professor R. H. Tawney took part in the deputation. Dr J. J. Mallon spoke at Stationers' Hall. The Earl of Ilchester took the chair at a meeting of the National Committee on 24 July (at which it was decided to set up a National Council for the Support and Defence of Books—the forerunner of the National Book League, soon to grow out of the National Book Council). In the House of Commons the speech on 25 July of A. P. (now Sir Alan) Herbert undoubtedly had great effect. Of the representatives of the Association, Geoffrey Faber's passionate belief in books, his eloquent powers of persuasion, and his sense of the occasion were invaluable and rightly drew the lights. Wren Howard's off-stage work of organization has not always received its due recognition. The considerable expenses of the campaign were met from a fighting fund previously collected from members of the Association by the foresight of the Treasurer and the Secretary. The removal of books from the category of goods whose production and sale were to be discouraged in war-time combined with the restricted scope of social life to bring about a remarkable revival in the home sale of books.

For the Forces. For the supply of books to the Forces collaboration by the War Office with an appeal launched by the Lord Mayor of London led to the creation of a Services Libraries and Books Fund. Since it confined itself to the purchase of second-hand books, remainders and sixpennies, took no notice of an offer to sanction the supply of books at best trade terms, and made no use of expert assistance, it cannot be said that it rose to its opportunities.

1941–1942

The year opened with renewed aerial bombardment during which immense quantities of books—estimated at more than 20 million—were destroyed in the London area. Many publishers suffered serious losses of stocks and not a few had to find new premises. In spite of these losses large stocks of unbound books still lay in warehouses throughout the country and as the year advanced and the demand for books grew by leaps and bounds, the diminishing capacity of the binding trade became a limiting factor. There were many pre-war books which until the end of the war publishers were unable to get bound or which came into and went out of stock as it became possible from time to time to bind limited quantities. As oranges became unobtainable the taste for them grew, and the same seemed to be true of books.

Labour. It was the binders who suffered most from the increasing depletion of their staffs as young men were called up and young women preferred to take employment protected by Essential Work Orders. The retention of younger women in publishing was somewhat easier. In July 1941, when it was announced that women in unprotected industries must be conscripted into the Forces and into war industries, it was agreed at any rate that publishing would be added to the list of industries whose female employees would not be considered for compulsory transfer to other industrial work and although the position worsened in November it was conceded that women under twenty and over twenty-five years of age would not be withdrawn without prior consultation between the Ministry and the Association.

For men—and the key workers in publishing and printing were generally older men—the Schedule of Reserved Occupations gave some protection. Although in March 1941 Sir William Beveridge, the Man-Power Adviser to the Government, had rejected a request from the Association that book-publishing should be classified as a 'protected' industry, in May after an interview between Lord Terrington of the Ministry of Labour and the Officers the Minister decided to give the trade a Committee of its own to advise the Ministry upon applications for deferment of calling-up from firms in book-publishing, book-printing and binding. The Ministry of Labour Book-Production Advisory Committee, as it was called, consisted of three representatives of the Association and one each of the printing and binding sections of the Federation of Master Printers.[1] It met for the first time on 26 May 1941 and before the end of the year had considered close on 1,000 applications, upon which its recommendations received the full support of the Ministry. The Committee remained in being until the end of the war.

Paper economy. An outstanding achievement of the year was the Book Production War Economy Agreement, which came into force on 1 January 1942. In the early summer of 1941 the paper situation became so serious that the Ministry of Supply called the attention of the Officers to the need for national economy and asked for an assurance that more economical use of the materials licensed for books would be made. The Officers expressed the opinion that with the support of the Ministry it would be possible to operate economy standards which would not unduly impair the physical appearance of books, and they were accordingly asked to present a plan. The Council therefore agreed in July to appoint a technical advisory committee,[2] and by October the committee had produced a schedule of standards and definitions. For new books it prescribed either minimum type-area to page-area ratios and maximum type sizes or a minimum words-per-page standard; and for new books and reprints it prescribed maximum paper substances and weights of binding boards. It took account also of the problems of special categories such as short books, books of poetry, and books for children. The novel, to take one standard example, was to contain not less than 375 words to the page instead of the normal 300, and the substance of the paper and the boards was to be reduced by about 20%. Books produced in conformity with the new standards were entitled to carry a colophon to that effect. Of the excellence of the committee's work the immediate and unquestioning approval of publishers and printers was the best evidence. In March signature of the War Economy Agreement, and no longer membership of the Export Group, became the necessary qualification for the full paper ration. It was only in some export markets, particularly in Canada in comparison with American books, that its inevitability was questioned.

[1] W. G. Harrap, G. C. Faber, R. J. L. Kingsford (the then Officers of the Association), Colonel O. V. Viney, Major R. (later Sir Robert) Leighton (for the B.F.M.P.).
[2] W. G. Taylor (Chairman), G. H. Bickers, W. A. R. Collins, G. F. J. Cumberlege (Oxford University Press), R. de la Mare, G. Wren Howard, R. A. Maynard (Harrap's), Stanley Morison (typographical adviser to Cambridge University Press and Monotype Corporation).

The Moberly Pool. During 1941 the paper quota had been reduced from 50% to 42% (for members of the Export Group). When at the December Council meeting the President announced that the ration was to be still further reduced in the New Year to 37½% (25% for non-signatories of the War Economy Agreement) he took some comfort from being able also to announce that at a meeting at the Ministry of Supply on 9 December[1] the Officers had been able to secure for the current period a pool of 250 tons, equivalent to an additional 2%, provided that the Association could satisfy the Ministry that it could be allocated in a satisfactory manner. Neither the Officers in initiating the proposal nor the Council in accepting the responsibility were aware of the storm which they were arousing. The pool was, primarily, to ensure the maintenance in print of books which were nationally important in war-time and the President anticipated that allocations from it were likely to be limited to reprints since the importance of reprints could more easily be established than of new books. Secondly, the Council was aware that applications to the Paper Controller for the printing of so-called essential books by people outside the publishing trade were increasing and were setting the Controller problems in the settlement of which he needed the shelter of professional advice. Nevertheless to some members of the Association the pool represented a potential threat of Government interference with the right of the individual publisher to decide what he was to publish and they were to watch its operation with suspicion until the end.

The Council accepted the President's nomination of four members[2] to form a Committee to receive the applications and advise the Paper Control. The Committee had had one meeting on its own and one with officials of the Control, to determine its procedure, when in February 1942 the requisite number of members requisitioned a General Meeting of the Association and gave notice of a motion opposing the pool as a development never previously discussed, as possibly objectionable in practice and certainly dangerous as a precedent, and resolving that the Ministry of Supply should be asked to receive a deputation and that the Officers of the Association should not make decisions on questions of principle

---

[1] At the same meeting, to the surprise of his fellow Officers, Geoffrey Faber's persuasiveness induced the Ministry to agree that if a quota-holder could find another quota-holder who was not using the whole or part of his quota, the former could take it over without deduction from his own quota.

[2] B. W. Fagan, R. H. Code Holland (Pitman's), R. J. L. Kingsford, W. G. Taylor. It may be said that they had experience in the publishing of learned, educational, medical, legal, religious and technical books, reprint series, and some general literature. Fagan, who was the first chairman and continued to be the *rapporteur*, was appointed C.B.E. in 1949.

without the prior approval of a majority of the members. The meeting was held on 3 March. The opposition to the Council was led, from different standpoints, by two powerful speakers, Douglas Jerrold (of Eyre and Spottiswoode) and Victor Gollancz; and the editor of the right-wing monthly, *The English Review*, and the founder of the Left Book Club may perhaps have been surprised to find themselves on the same side. Gollancz's opposition was to acceptance of the principle that someone was going to decide which books were essential and which were not—who could decide the relative importance of a Roman Catholic Breviary and an edition of Karl Marx?—and the proposer and seconder of the original motion withdrew it in favour of one by him expressing the wish that all available paper should be allocated to publishers on a percentile basis for their free use. But there was plenty of support for the pool and when Bertram Christian, doubting whether the vote on the amended motion would settle anything, moved 'the previous question' it was carried by 48 votes to 28. The Annual General Meeting of the Association was to come on 26 March and before that day G. Wren Howard gave notice of the following motion:

> This meeting approves and endorses the Council's action in agreeing to assist and advise the Paper Control in the allocation to members of small supplementary supplies of paper additional to the prevailing quota. It recognizes that no question of censorship is involved or implied by such an arrangement but that the only object is to mitigate some of the inequalities in the distribution of paper inevitably attendant upon the adoption of an arbitrarily chosen reference period as basis for the paper quota. Finally it records its desire that should any threat of censorship of books become apparent at any time a special general meeting of the Association shall immediately be convened so that the matter may be discussed by the membership.

After some opposition, on the grounds that censorship was involved and that, if it was inevitable, it had better be done by a Government Department than by a committee of publishers, the motion was carried by 62 votes to 11.

Although the pool was to come under criticism again as the tonnage of paper allocated to it for so-called 'essential' books was increased and when in September 1947 a separate pool to assist exports was formed, it continued to exist until January 1949. The four publisher members of the advisory committee remained unchanged, with the addition of an independent chairman, Sir Walter Moberly,[1] in December 1942; and although

---

[1] then chairman of the University Grants Committee, formerly Vice-Chancellor of Manchester University.

Professor (later Sir Arnold) Plant,[1] became chairman in July 1945, it was to be 'the Moberly Pool' until the end. The quantity of paper in the pool, which was at the rate of 1,000 tons p.a. at the beginning, was increased to 5,850 tons p.a. in 1948; and the export pool of 750 tons p.a. in September 1947 was increased to 3,000 tons p.a. in 1948. During the seven years of its existence the committee held 86 meetings, considered some 10,300 applications (which followed them on week-ends and on holidays, when holidays became possible), and allotted 23,700 tons of paper. Having been at first advisory to the Ministry of Supply, the Committee became responsible to the Board of Trade in December 1942 when that Department, as will be recorded, assumed responsibility for books. Of the civil servants associated with the committee pleasant memories survive of two in particular, T. Brown of the Paper Control and C. M. P. Brown of the Board of Trade, known inevitably as Brown Paper and Brown Board.

What of the fears of Gollancz and Jerrold? The Ministry of Supply wisely refrained from a definition of 'essential books', and the committee got no nearer to it than a definition of the inessential book as 'the book I don't happen to want to-day'. In the minds of the Committee the essential book did not exist *in vacuo* but only in the context of other applications; and if an application was unsuccessful in one period, it might be successful in another, or the publisher could print the book from his own quota and apply for another book in the next period. It may have seemed an inquisition to publishers who had of course to reveal the use of their quotas, but the fears of censorship were unfounded. Reduction was more common than rejection: in 1942 one-third of the amounts applied for was allotted, rising in 1947, when there was more paper in the pool, to two-thirds. Up to the end of the war the paper allotted fell into the following categories (in percentages):

Educational	25·4
Medical	17·4
Scientific and Technical	15·8
Religious	15·2
General	13·1
Dictionaries	10·0
Legal	3·1
	100·0

Cloth and boards and salvage. During the year 1941–2 stocks of cloth and binding boards were used up and the continued production of books in cloth boards seemed precarious, but by the end of the year adequate supplies seemed to be assured. It had been necessary to convince the

[1] Professor of Commerce at the London School of Economics, 1930–65.

Ministry of Supply that it would be impossible to go over to the continental practice of paper covers, because the different machines were not available in sufficient quantities and additional labour for hand work would be needed.

Realizing that supplies of paper and boards depended in part upon the collection of salvage, the indefatigable President put forward in September 1941 to the Controller of Salvage a scheme for the collection of unwanted books and the extraction of those valuable to libraries and those suitable for the Forces before the remainder went for salvage. From this grew the National Book Recovery Appeal, organized by a Central Council on which the Association was represented, with Book Drives and local scrutiny Committees. By February 1943 7,620,030 books had been recovered, from which 150,000 had been reserved for libraries and 704,000 had been dispatched to the Forces.

The year 1941 also brought into being the War Damage Act, Part III, covering the insurance of publishers' plant and requiring the settlement with the Federation of Master Printers of problems of the dual ownership of standing type and the doubtful ownership of work in progress and photolitho negatives. The Central Price Regulation Committee also came into being, with its local committees also operating the Location of Retail Businesses Order 1941[1] and occasionally crossing the work of the book-trade's Joint Advisory Committee.

1942–1943

Towards the end of 1941 the Officers and the Council had made renewed efforts to obtain protection for the book industry in some acceptable form and co-ordination of its needs. If more paper were made available, would there be men and metal for the printing, women and boards for the binding, string for packing, and shipping space to New Zealand? In October 1941 Viscount Samuel had opened a debate on book-production in the House of Lords; and when Brendan Bracken became Minister of Information, the President (Walter Harrap) had sought an interview with him, as a result of which the Officers were called to a conference on 11 November of officials of three Ministries, H.M.S.O., and the British Council. The conference had produced no worthwhile result, nor had a deputation, led in December by the Treasurer, to the Board of Education, to request in particular its support for an Essential Work Order for the bookbinding industry.

[1] see p. 198.

Board of Trade as sponsors. It was December 1942 before 'protection' was achieved, and when it came the Council was quick to see that the conditions attached might involve some further loss of individuality—as the institution of the Moberly Pool had done—and that in the necessary readjustments solidarity for the common good of the membership was more than ever necessary. With the diminishing accent on exports, as American Lend-Lease reduced immediate problems of payment, complaints grew throughout 1942 that books needed at home for educational and other services were often unobtainable and led to the appointment by the Lord President of the Council of an inter-departmental committee under the chairmanship of Osbert Peake, M.P., which recommended that books should have a sponsoring department. The responsibility was assigned to the Board of Trade, with these terms of reference: 'Book production itself will henceforth be treated as production of national importance, and it will be the responsibility of the Government to ensure that labour and materials are available for the quantities and kinds of books which it is deemed in the national interest to produce.' In December the Officers were summoned to meet the President of the Board of Trade, Dr Hugh Dalton. He was to prove an unreliable godfather to publishers; and the Association's President in 1943–5 was to learn that the parliamentary question could get a reply from which *suppressio veri*, or even *suggestio falsi*, were not absent.

In their first meeting with Richard Pares, the Assistant Secretary at the Board of Trade to whose charge books had been assigned, the Officers were asked to consider how five categories of books—educational, technical, medical, cheaply priced children's books, cheap series of a popular nature—could be produced in larger quantities. At a special meeting in February 1943 the Council considered such solutions as the redistribution of paper quotas according to categories of book or categories of publisher, only to dismiss them in the belief that the needed categories would change, that there was already as great a shortage of scientific books, religious books, and dictionaries, and that the emphasis should not be on certain categories but on good books in every category. There was no answer which did not involve the allocation of more paper for books.[1] But the ration remained at $37\frac{1}{2}\%$ throughout the year.

Economy standards. The standards imposed by the War Economy Agreement continued to receive general acquiescence and by February 1943

---

[1] The Council's examination of the problems was conveyed to the Board of Trade in a report dated 22 February 1943.

the agreement had been signed by 750 firms, persons and institutions in receipt of paper quotas, of whom the total membership of the Association comprised only 162. In November 1942 a Paper Economy Committee of the Ministry of Production suggested that book-jackets should be abolished, but the case made for their retention was accepted. It may be noted that many publishers were then using the back flap of suitable jackets to publicize the overseas broadcasts of the B.B.C.

Metal. As the year advanced the shortage of metal for type, blocks and plates became acute. Zinc for the making of new line blocks after July 1942 became entirely dependent upon the salvage of old blocks. In January 1943 it was calculated that since the beginning of the war members of the Association had released for distribution some 5,000 tons of type metal, representing perhaps 11,000 titles, but the Association then appealed to its members for yet more.

Men and Women. Although materials became increasingly scarce, it was the shortage of labour which caused the gravest concern. The Book Production Advisory Committee of the Ministry of Labour was able to secure the retention of essential men of military age within the publishing, printing and binding trades, but was not permitted to cover men over 40, who were liable to transfer to other industries. Attempts to extend the scope of the Advisory Committee to include women coming under the National Service Acts also failed, but the Association was at least able to make recommendations to the relevant Man-Power Boards through the Board of Trade's Deferment Officers.

1943-1944

Although there was some increase in the supply of paper the problems of using it were made more difficult by the absolute priority exercised by H.M. Stationery Office over book-printers and binders. Existing demands raised the turnover of the trade from £9,950,000 in 1940 to £19,300,000 in 1943—and the Price Control Act and excess profits tax ensured that there was little or no increase in the level of prices—and the year brought evidence of coming new demands. The London County Council and other education authorities were seeking replenishment of the books which they had been using since the beginning of the war and the new Education Act emphasized the inability of publishers to supply the existing needs of schools and, still less, the books which a more progressive educational policy would require. To the immediate needs of the Forces for recreational reading were added the coming requirements of the Army Post-War Education Scheme. The needs of colonial schools were represented to the Association by the Colonial Office. As the liberation of Europe began, its hunger for books was a major concern of the Conference of Allied Ministers of Education; and finally, by a swing of the pendulum, all exports resumed the importance which they had held in 1940. In 1943 the export percentage of the book trade had fallen to 23%, as against 35% in 1940. There were not enough books to go round[1] and by the spring of 1944 it was estimated that nine out of ten new books were oversubscribed before publication.

Paper. The proposals with which, as we have seen, the President of the Board of Trade began his sponsorship—that more paper should be put into five categories of books at the expense of other categories—he finally reduced to one category, 'cheap books'. The Council repeatedly emphasized that necessary books existed in all categories and that it had no authority from members of the Association to discuss discriminatory proposals and could only do so under compulsion. Nevertheless the Board felt bound to pursue its intention, in the form of a proposal to reduce the rate of the quota from $37\frac{1}{2}$% to 35% and for the use of the resulting 1,300 tons p.a. to form a new pool of 1,000 tons for cheap books (up to 5*s*. in price) and to increase the Moberly Pool by 300 tons. Thereupon the Council made arrangements for the calling of a special General Meeting of the Association, but did not proceed with them when the Board agreed that before its proposal was put into effect, information should be collected about its probable effect. Upon examination of statistics collected from members the Board became convinced that sufficient paper was already being put into cheap books and therefore dropped its proposal. In September 1943 the allocation of paper for newspapers and periodicals was increased by an amount considerably greater than the total allocation for books, and the Council renewed its pressure. In reply to questions concurrently asked in the House of Commons the Minister of Production, Mr Oliver Lyttleton,[2] gave the tonnage of paper being allocated to newspapers, the Stationery Office, and books as: newspapers 250,000 tons, Stationery Office 100,000 (including 25,000 for the War Office), books 20,000; and in November he announced that he proposed to make additional supplies available for books. Knowing

---

[1] 'The Out-of-print Book' was the subject of an article contributed by the Treasurer, B. W. Fagan, to *The Times Literary Supplement* of 7 August 1943.
[2] later Lord Chandos.

as it did the attitude of the Association to any proposal which would reduce the individual publisher's freedom of publication, the Board of Trade did not discuss the method to be adopted for the distribution of the additional supplies, but the Council did not feel obliged to oppose the addition of 700 tons to the Moberly Pool accompanied as it was by the addition of $2\frac{1}{2}$% (1,300 tons) to the quota, raising it to 40%.

In June 1943 it was ruled by the Ministry of Supply that Directories, Year Books and similar publications did not need the full ration of paper and would in future be entitled to 25% only. The Association was able to secure the reinstatement of annual publications, whether described as Year Books or not, of which the contents were new each year; and it also appeared that publishers of annual publications for which 25% only was still licensed were not prohibited from supplementing it with unlicensed paper, although the acquisition of any large amount in this way might lead to a retaliatory prohibition.

'Free' paper. Here was an example of the anomaly of the so-called 'free' paper which printers could draw under a general licence, to enable them to undertake miscellaneous printing, perhaps of a political nature, to which otherwise the Ministry of Supply would have been in the invidious position of giving or withholding a specific licence. From October 1943 the Council was continually pressing the Ministry of Supply to end the injustice to quota-holders of permitting the unlimited use of 'free' paper by firms without quotas; it urged without success that the use of such paper for books should be prohibited or the embargo on its use by quota-holders should be lifted. Evidence was provided of the activities of the new publishing firms relying on the 'free' paper (which by January 1944 amounted to 120 in number), of the damage to established firms through loss of their authors, and of the poor quality and high prices which some of the newcomers were able to get away with in the general shortage of books. In its protests the Council had much support from questions in the House of Commons and in the press. *The Times* in a leading article on 21 October 1943 attacked the anomaly that 'a new publishing business may start without restrictions while at the same time a new periodical-publishing business, or a new retail business, requires a permit'; and it gave its support to a letter from the President of the Association which appeared in the same issue. In his letter the President argued that if catch-as-catch-can was to be the game, the established firms as well as the newcomers should be permitted to play it; and he went on to defend the Association's refusal to discriminate between one kind of book as 'essential'

and another kind as 'inessential'. 'The essential book' he wrote, 'is the book which we all happen to want to-day; and, as it was put in a recent leading article in *The Times Literary Supplement*, a new definition of "tripe" might well be "the book which I do not want to read myself".' The anomaly was also evident in the insistence by the Council that if, as the Paper Economy Committee of the Ministry of Production proposed, the substance of the papers used for book-jackets was to be limited, then all the producers of books and not merely the quota-holding signatories of the War Economy Agreement should be required to observe this limitation, and indeed all the other limitations enforced by the Agreement. To this the Ministry of Supply would not agree and accordingly the limitation on book-jackets was not put into force.

The Association also tried unsuccessfully to right an injustice in the supply of paper towards the replacement of stocks lost in air-raids. Those publishers who were 'blitzed' in 1940 received compassionate allocations, those hit after mid-1942 received an additional ration up to $7\frac{1}{2}\%$ of the reference period, but those who suffered in the severe raids of the summer of 1941 received nothing.

Discriminatory pressure. Although it was the declared policy of the Association not to support differentiation between one kind of book and another, the Council was not without pressure from two of the Groups to permit them to make their own case for preferential allocations of paper for their own books. In July 1943 the Medical Group, and in March 1944 the Educational Group, were informed that memoranda which they had prepared advocating the special needs for more medical and school books should not be sent to any Government Department, but with some alteration might be sent to influential people and professional bodies as examples of the general shortage of books.

The small publisher. To the special difficulties of small publishers, for whom F. J. Warburg of Secker and Warburg was a principal spokesman, the Council could be more sympathetic. The smallness of their paper quotas made it difficult for them even to stay in business and the prospect that one of their books might become a best-seller was almost a nightmare. A memorandum from seven small firms was submitted to the Council in February 1944, and although it seemed inevitable that if more were given to the small it would be taken away from the large and that *bona fide* small publishers must somehow be defined and sorted out of some 450 users of paper holding small quotas, nevertheless in October the Officers put up to the Ministry of Supply a scheme under which small independent firms engaged solely in book-publishing would have a basic 5 tons of paper, to which their individual quotas would be added, subject to a maximum of 20 tons per annum. But there was to be no concession until May 1945 when the Ministries of Supply and Production agreed to the provision of a new pool of 225 tons of paper per annum, from which on the advice of a new committee grants would be made which would bring the quotas of small publishers up to 75% of their pre-war usage, with a maximum of 20 tons. The committee[1] began to operate in September 1945, but the rapid subsequent rise of the ordinary quota made it short-lived.

Labour. At the end of March 1943 the Officers were informed by the Board of Trade that, in consequence of the shortage of men and women for war industries in certain regions, it was proposed to prohibit the continuance of printing and binding in those areas. By July the proposal had been abandoned, but the three sections of the book-production industry were to be given a fresh squeeze by a new panel, advisory to the Ministry of Labour and consisting of representatives of the Association,[2] the Federation of Master Printers, and the printing trade unions. Postponement was secured of the review of publishing staffs and of women in binding, where already machines were standing idle for lack of hands, but the review of men in printing went forward.

The Forces. During this year official help was at last forthcoming for the much-needed supply of books for the recreational reading of the Armed Forces. Additional paper was granted to the War Office for the production of Guild Books,[3] Penguin Books, and other cheap paper-bound editions and the Services Central Book Depot was occasionally able to supply paper for small numbers of suitable books to be 'run on' at the time of a general reprint. In September 1943 the Association became aware of the size of the coming programme for the Services Post-War Education Scheme, which was being planned by the Director-General of Army

---

[1] B. W. Fagan, R. J. L. Kingsford, W. G. Harrap, Bertram Christian.

[2] The Association's representatives were the same as its representatives on the Ministry of Labour's Book-Production Advisory Committee (G. C. Faber, W. G. Harrap, R. J. L. Kingsford, B. W. Fagan).

[3] Just before the war, when publishers were showing some reluctance to license Penguin reprints competing with their own cloth-bound reprint series, Walter Harrap had initiated the British Publishers' Guild, a co-operative organization through which any member of the Publishers Association who wished to join could finance and market paper-covered reprints of his publications.

Education, Philip Morris.[1] Some three million books, consuming 1,000 tons of paper, were to be ordered by H.M. Stationery Office and the priority exercised over printers and binders by that very department was to be the greatest obstacle to their production.

Europe. To prepare for the post-war world there also came into being early in 1943 the Conference of Ministers of Education of the Allied Governments in the United Kingdom, no less than nine in number: Belgium, Czechoslovakia, Greece, Luxembourg, the Netherlands, Norway, Poland, Yugoslavia and the French National Committee. The Conference set up a Books and Periodicals Commission, under the chairmanship of Professor (later Sir) Ernest Barker. It may be of interest to note that in its wish to aid the use of English, instead of German, books in European universities the Commission called attention to the difficulty that in certain subjects English books did not use the metric system. The change was to wait for more than twenty-five years and in 1943 the Council could do no better than follow the President's suggestion and recommend publishers to print metric equivalents in brackets after the English measurements.

1944-1945

Throughout the year it was a nice question whether more paper or more labour was the greater necessity; and it seemed that the Board of Trade as the sponsor for books accepted the shortage of one as a sufficient reason for not trying to get more of the other. As classics, dictionaries, and standard works were got back into print, the standing orders were rationed and they then went out of print again. More paper, by permitting longer runs, would have enabled the existing labour to be used with greater economy; and the inadequacy of the available labour was increased by the continual call made upon it by H.M. Stationery Office. The rapid improvement of the military situation in the autumn of 1944 made it not unreasonable and not unpatriotic to press for some release of printers from civil defence and munitions and some direction of younger women into binding; and to these demands, in relation to the Services Post-War Education Scheme, the Stationery Office gave effective support.

The parliamentary deputation. In May 1944 statistics of stocks, of books printed and bound during the last year, and of machines standing idle were incorporated in a 'Report on the state of the book-publishing trade in

---

[1] Later Sir Philip Morris, Vice-Chancellor of Bristol University; previously Director of Education for Kent.

relation to coming demands'. It was widely distributed and favourably received by the press, which throughout the war gave unselfish support to the case for more paper for books, and by leaders of parliamentary and economic opinion; and in subsequent months it was supplemented by the preparation of documentary evidence for the use of members of both Houses of Parliament in a growing volume of criticism of the Government's policy towards books. This criticism culminated on 21 December 1944 in the reception by four Ministers (Production, Supply, Labour, Board of Trade) of an all-party deputation of members of both Houses, led by Viscount Samuel and H. Graham White, M.P., which pressed for a substantial additional allocation of paper and the release of labour. Only the Minister of Labour, Ernest Bevin, gave an indication that he might do something, and he was as good as his word. It might have been expected that the President of the Board of Trade, as the official sponsor of books, would at least have given a fair wind to the arguments of the deputation, but Dr Dalton launched an attack on publishers for wasting their quotas on 'worthless' books—and this came from the Minister who the year before had proposed discrimination in favour of cheap books and was permitting 'mushroom' firms to spring up on 'free' paper. To this criticism Dr Dalton added some praise of the Moberly Committee and, fulsomely, of the contribution to it of the President of the Association, who had accompanied the deputation as adviser and was sitting in frustrated silence at the back of the room.

Quota and pool. In October 1944 the paper quota had been raised from 40% to $42\frac{1}{2}$% and the Moberly Pool had been increased from 1,700 to 2,300 tons p.a., which brought the total allocation of paper for books to approximately 24,500 tons a year. The smallness of the increase, intended particularly for export to Europe, was obscured by Dr Dalton's answers to parliamentary questions in the House of Commons, which led members to suppose that the total allocation had been increased to 40,000 tons and that the supply of school books at home would be materially improved.[1] In November also the President had an interview with the Minister of Education, R. A. (now Lord) Butler, and found him and his officials not unnaturally misinformed about the paper available for books for schools.

In January 1945 the Pool was increased, as the result of the parliamentary deputation, by a further 1,000 tons without any increase in the quota; and, as it was bound to do, the Council called a General Meeting

---

[1] Dr Dalton's inaccuracy was the subject of a letter from the President to *The Times* on 19 October 1944.

of members on 7 February. At the meeting it was evident that the Pool was acceptable to most members, provided that it remained relatively small and that the Committee remained free from Government interference, but it was agreed that any further increase in the proportion which the Pool bore to the total of individual quotas would be highly objectionable.

'Free' paper again. During the year the Council persisted in its attempt to end the injustice of the 'free' paper situation and, as the result of questions asked in the House of Lords by Lord Elton, the Officers were able in April and July 1944 to take the Association's case up to the Minister of Supply, Sir Andrew Duncan. That supplies of 'free' paper then began to diminish was to be deduced from the announcement in December of the formation of a new association, the British Federation of Book Publishers, with the advertised object of obtaining paper quotas for new firms. The attitude of the Council was that quotas should not be established for newcomers while pre-war firms were still subject to severe restrictions.

Electoral registers. Every demand of the Association for more paper required justification of the industry's ability to print and bind it; and the position seemed likely to be made more difficult by the printing of the new Parliamentary Registers for the coming general election. But there was some improvement by the end of the year. Deferment of military service granted to men born in or before 1909 was continued indefinitely; the Stationery Office secured the return of some men to assist in printing the books required for the Services Post-War Education scheme; and of the 900 additional operatives required for the printing of the Parliamentary Registers at any rate 700 had been provided by the end of February 1945.

1945–1946

Within the space of four swift months there came the fall of Germany, and of Japan, the end of the Coalition Government, the advent of the Labour Party to power, and the end of Lend-Lease. The Association was brought face to face with a new set of problems—of reconversion to the shape of the post-war world—with which this chapter will not be substantially concerned, but war-time shortages also persisted.

Paper. The end of the war brought no immediate abatement in the demand for books, and to it was added the now urgent need to export more and more. The paper situation steadily improved. The quota was increased in

July 1945 from 42½ to 50%, in November from 50 to 65%, and in March 1946 from 60 to 75%. The tonnage allocated to the Moberly Pool remained unchanged. In March also an additional 10% was made available to all quota-holders under an export target scheme agreed between the Officers and the Board of Trade. The Export Group, whose exports for the twelve months to 30 June 1945 totalled about £5,000,000, undertook to try to reach an annual total of £8,000,000 by the middle of 1947. The additional paper was not tied to particular orders; each firm assessed, on a generally assigned basis, what its own contribution should be or was free to forgo the extra allocation; and there was no inquisition and no penalties.[1]

During 1945 the easing of the paper position enabled the Ministry of Supply to grant small quotas to new publishers, particularly to ex-servicemen, to make allowances to some publishers who had not had them in respect of losses by enemy action, and to raise the quotas of small publishers to 100% in cases of special hardship. Additional paper was granted for Directories and Year Books and for catalogues and other advertising materials; and finally the publication of guide books, which had gone from the market when the signposts were removed from the roads, became permissible again.

Economy standards ending. Of the additional supplies of paper some had inevitably to be consumed in a partial relaxation of the maximum type sizes, areas of margins, and paper substances enforced by the War Economy Agreement. Although Canada in particular urged the complete and immediate abolition of the economy standards, it was felt that it must be done in two stages, and to the first the Ministry of Supply agreed in October 1945. The permitted weights of binding boards were also raised in April 1945.

Labour. Shortage of paper, then, the most constant of the war-time shortages, had given way to shortage of man-power. The seriousness of the labour prospect was shown by an estimate of the Federation of Master Printers in the autumn of 1945 that even when demobilization and transfer from war industries were complete the printing and binding trades would be some 10,000 skilled operatives short of the 1939 figure of 70,000. At the same time it also appeared from statistics collected by the Association from

[1] In spite of the fuel and power crisis of February 1947, which caused a reduction in the paper quota and production difficulties of all kinds, the export total for 1947 was approximately £7,500,000.

its members that there were, apart from current production, 43,000 titles lapsed during the war which it was intended to put back into print, either as reprints or in revised editions, and 9,000 new books awaiting production in authors', publishers' or printers' hands—a total of 52,000 titles waiting in a production queue. In February 1946 the accelerated release of 750 printing operatives under Class B of the Demobilization Plan was announced in the House of Commons.

10

1939-1946 (2)
World War II: the export markets, trade relations at home

THE EXPORT MARKETS

Before the war the book trade's visible exports had averaged 30% in value of its total sales. Throughout the war in spite of shortages of labour and materials and of the knowledge that most of the books which were produced could be sold several times over at home, in spite of shipping difficulties, and in spite of the Government's diminished emphasis on exports after the beginning of American Lend-Lease, the percentage never fell below 21 and recovered to 23·4 in 1945. Whether the national economy needed exports or not, the Association never forgot that for the book trade the home market by itself was insufficient. Europe was closed, the Middle East and the Far East virtually closed. There remained the countries of the Commonwealth, in which British publishers were accustomed to have an exclusive market for books of British origin and for British editions of books of American origin, and the countries of South and Central America, where they expected to retain a share in an open market. Over the Commonwealth there was some disagreement between British and American publishers, and their respective associations, which was distasteful as between two allies, and as the war came towards its end, some suspicion that ambassadors of the American way of life were also doing some profitable trade in books. But transatlantic visits by representatives of the two associations helped towards an understanding.

Book Export Scheme. Early in 1940 the British Council completed a well devised scheme to aid the foreign sale of British books of cultural value. The Book Export Scheme—B.E.S. as it came to be known—was planned by a committee under Stanley Unwin's chairmanship, working in collaboration with the Association and with the backing of the Export Credits Department of the Board of Trade. It provided that selected booksellers could order 'on sale or return'; that if at the end of six months the books had not been sold the publisher would credit them in full; that the bookseller would then hand them over to the local representative of the British Council, to which they would be recharged at half price less $2\frac{1}{2}\%$; and that the books would then be used by the British Council for presentation purposes. Payment for the books sold by the booksellers was guaranteed to the publisher in certain circumstances, for example currency restrictions, by the Export Credits Department.

Although B.E.S. was temporarily left in the air by the disappearance of the European market at which it was principally aimed, by the end of 1942 it was being operated successfully in other countries (particularly in South America), to which also more than 200 books a month were being sent under an accompanying scheme for the distribution of review copies to selected periodicals. When in 1944 the British Council was designated by the Government as the channel for the supply of books to the liberated countries, B.E.S. Ltd was established with three representatives of the Association on its Board,[1] to undertake the trade between British publishers and foreign booksellers until direct trade should be possible.

South America. Early in 1940 the Association became aware that American publishers, stimulated by the State Department, were driving hard in the South American market. The British Council then began to increase its activities in that continent and in 1941 put forward plans for the maintenance there of a permanent representative of the book trade to be financed jointly by the British Council and British publishers. In March 1942 the first representative, F. M. Scully, was sent out and the joint agreement continued throughout the war.

The liberation of Europe. As the liberation of Northern Europe began, the declared aim of the Association was to achieve as quickly as possible direct trade between British publishers and exporters and the foreign importer and to accept official intervention only so long as difficulties of transport and currency exchange could not otherwise be overcome. B.E.S. Ltd was the accepted channel for the sale of books in the English language; the placing of translation rights with approved publishers in the liberated countries and in Germany was undertaken until June 1945 by the Supreme Headquarters of the Allied Expeditionary Force (SHAEF), in which Captain Spencer Curtis Brown, widely known in peace time as a literary agent, was the responsible officer, and subsequently by the

[1] Stanley Unwin, W. G. Harrap, R. H. Code Holland.

Ministry of Information. Conditions were different in each country and subject to constant changes. Denmark provided a model of smooth working: immediately upon liberation the Government produced a list of non-collaborationist publishers and booksellers and guaranteed them import licences and sterling exchange for payment. For France an intergovernmental scheme for the expenditure of £100,000 on British books involved complicated official procedure on both sides, the French orders requiring approval by the Bureau Professionel du Livre and transport and payment being effected through B.E.S. Ltd. For Belgium and Holland the selection and buying was done in bulk by a Government agent and individual books required by libraries, universities and medical schools could not at first be paid for. But by February 1946 direct trading was in operation with Belgium, Holland, Denmark, Norway, Sweden, Finland, Spain, Portugal and Switzerland. In July 1945 the Swiss book trade had been the first to send a delegation of publishers and booksellers to London.

American competition. As shortages of materials, labour and shipping bore more hardly on Britain than on the United States with her greater internal resources, American publishers saw openings to be seized in countries which British publishers had regarded as their preserves and indeed saw a positive duty to satisfy a hunger for books if British publishers could not do so. Some American publishers also saw an idealistic mission in helping these countries to rid themselves of the remnants of the oppressive British colonial system. In the exchanges which followed there were two sources of misunderstanding: first, that there were many American books which did not find a British publisher and which the American publisher would expect to sell direct to, for example, Australia; secondly, that it was not just the British Commonwealth by constitution which British publishers expected to keep in their contracts for books to be published in both Britain and the United States, but rather the traditional British market, including for example Palestine and the Middle East, which they had cultivated before the war. If the stakes were greatest in Australia, Palestine became a sort of symbol of the principle at issue and in 1944 the Council of the Association was expressing its increasing concern at the sale there of titles in the American Pocket Books series in infringement of British publishers' market rights. Of the countries which were temporarily starved of books in the English language India[1] was perhaps the one to which America felt the greatest idealistic call.

[1] Import controls, conditioned by the available shipping, reduced shipments of books from Britain to India to 86 tons in the second half of 1943, to be compared with 800 tons in the following full year.

It was the English-speaking countries which gave the Association most cause for concern. In South Africa the increasing importation in 1944 of American editions of British copyright books required the Council to organize action by the local representatives of British publishers under the leadership of H. B. Timmins, and in 1945 the representatives formed themselves into an advisory committee on the lines of the one in India. In New Zealand it was proposed in 1943 that, if the necessary import licences[1] could be obtained, it should be permissible to buy American editions of books of which the British editions were unobtainable. Here and elsewhere the Council, at the risk of being thought a dog-in-the-manger, felt bound to maintain the principle of the indivisibility of the British market. The situation was not helped by the preference which was felt for the physical appearance of the American book to the British book produced under the war economy standards; and British publishers had something to learn from the greater effectiveness in some markets of American book-jackets and promotional material. It was in Australia, with its close partnership with the United States in the Far Eastern war, and in Canada, with its natural affinity with its southern neighbour, that these problems were most acute.

Canada. The problem of keeping British books in Canada was in part not a new one; it was a market towards which before the war British publishers had been short-sighted and somewhat defeatist. There were but eight million English-speaking inhabitants scattered over a territory larger than the United States, giving a small market for books, expensive to reach, in which publishing houses, whether Canadian or British, could get an adequate volume of business only by each undertaking an agency for many lists, American as well as British. In July 1939 a report from H.M. Trade Commissioner in Toronto called attention to the decline in Canadian imports of British as compared with American books and analysed the causes: the predominance of American periodicals reviewing American books, Canadian preference for the larger format of, for example, American novels, the effect of the Canadian climate, except in British Columbia, on British bindings, and the better salesmanship of American jackets. To these handicaps were added the delays to which transatlantic shipping in particular was subject in the autumn of 1939 and 1940 and the sinking of three large shiploads of books shortly before Christmas 1942. It was then in a Canadian book trade almost denuded of British books

---

[1] In 1940 New Zealand's balance of trade had caused her to reduce some imports, including books, by 50%.

The export markets 191

that three delegates of the Association arrived, as will be seen, in May 1943.

Australia. If British publishers were somewhat half-hearted in Canada, it was not so in Australia. Before the outbreak of the war members of the Association, with a few exceptions, had bound themselves not to enter into a contract for any book from which the Australian and the New Zealand markets were excluded, with the exception only of books by authors resident in either Dominion.[1] The determination to hold on was strengthened in 1941 when two events persuaded the few important firms who had not signed the agreement to add their signatures. In that year the Defence (Finance) Regulations governing payments to non-sterling areas at first prohibited, and subsequently limited to £50, the payment of advances, on account of royalties, to American authors and it seemed that Australian publishers would seek to buy Australasian rights in American books direct, until it became known that the Australian exchange control authorities were applying the same limitation. Secondly, Australian imports from outside the sterling area became subject to control and in April 1941 it was realized that books printed in the U.S.A. even though imported from the United Kingdom would require licences. It was conceded, however, that a certain degree of manufacture in the U.K. might give exemption and that the British binding of American sheets would probably be sufficient.[2] One other consequence of Government action must also be mentioned. In 1943 price regulations were introduced in Australia and the Australian Booksellers Association secured approval of its schedule of retail prices by the Prices Commissioner. This schedule, like previous attempts, was based not on the English published prices, with a mark-up to cover transportation and exchange, but on the trade discount, and so discriminated between publishers according to their terms and discriminated against educational books, on which smaller discounts were given than on general books.[3]

Hitchcock's mission. Australia and Canada were to become the principal subjects for discussion in a series of transatlantic missions. At the end of October 1942 the Association was pleased to welcome Curtice Hitchcock,

[1] see p. 121.
[2] In 1944 the Import Licensing Department in the U.K. had to be convinced of the importance of imports for re-export if the Commonwealth market were to be retained. It finally agreed in December to license, under a scheme proposed by the Association, additional imports for re-export up to £20,000 p.a. which the Treasury had undertaken to release.
[3] The Australian schedule is still subject to the same criticisms today (1970).

President of Reynal and Hitchcock, who came as the official delegate of the American Book Publishers Bureau and of the United States Office of War Information. The purpose of the mission, as defined by Elmer Davis, Director of the O.W.I., was:

(1) to convey to the British Book Trade, which has suffered severe hardships in this war, messages of friendship and goodwill from the American Book Trade, which has not been permitted to send a representative since 1939; (2) to observe at first hand British war-time publishing conditions, so that the American trade may benefit from its experience; (3) to discover how Anglo-American relations may be extended and improved through an exchange of ideas through books.

Hitchcock's arrival was celebrated on 3 November at a dinner given by the Association and the Associated Booksellers and held, by courtesy of the Lord Mayor, in the Mansion House; and his visit was to last for almost two months. At the November meeting of the Council, of which he was elected an honorary member for that meeting, he went directly to the problems of the Australian market. He asked, first, whether it would be possible for those books which British publishers could not, owing to present difficulties, supply to Australia to be supplied from the U.S.A. Secondly, he urged that British publishers should modify their present policy towards Australian publishing rights in books of American origin so that American publishers could dispose of these rights independently of other British publishing rights; and he stressed the increasing affinity between the U.S.A. and Australia and the claims of the local Australian publishing industry for recognition. To the first suggestion the Council replied—perhaps with some relief that the answer was taken out of their hands—that Australian imports from the U.S.A. were limited by currency restrictions. As to Australian publishing rights the answer was that there was as yet no Australian publishing industry. With a few exceptions Australian publishing was undertaken by Australian houses of British publishers; and the exceptions were primarily retail booksellers so that a sale of rights to one of them was liable to be suspected of giving a near monopoly and preferential terms to one bookseller, to the strong disapproval, even if unjustified, of the rest of the Australian bookselling trade. Moreover, if British publishers, especially those with Australian houses, were to be expected to give up Australian rights in the best-sellers, they would be unable to market the many other American books of merit the sales of which in Australia were relatively small. After the meeting Hitchcock was given a reasoned statement for presentation to his Bureau; its standpoint is clearly expressed in the following extract:

The export markets

While British publishers have a wide market for the books they issue they are able to undertake the publication in this country of a large variety of American books. Take away these overseas markets and it will be found that many American books will fail to find a British publisher. The market for books in the English language is not yet large enough in Africa, India, or any of the Dominions to make possible the institution of local publishing industries. Many of the English-speaking countries overseas must rely on Great Britain for their supplies of English books. Publication in Great Britain usually means that there is a wider distribution of books throughout the whole world than could be achieved if the books were only published in the United States.

The time was to come, nevertheless, when some American publishers would not publish only in the United States, but would find it possible to establish branch houses of their own in, for example, Australia.

The Association's mission. Curtice Hitchcock's visit was quickly followed by an invitation from the American Book Publishers Bureau to the Association to send three representatives on a return visit, and the invitation received the support of the U.S. Office of War Information, which was to make it possible for the delegates to see something of the American war effort, and of our Ministry of Information, without which their transatlantic passage would have been impossible. It was imperative also that they should visit Canada and an invitation from the publishers of Canada was quickly forthcoming. The three immediate past Presidents—Walter Harrap, Geoffrey Faber and G. Wren Howard—were nominated and the party landed in New York from one of H.M. troopships on 13 April 1943.

In New York. In the discussions which ensued in New York about the exclusive territories which the British and the American publisher would normally expect to have, the delegates relied on the statement which had been given to Hitchcock, supplemented by what came to be known as the 70:30 plan. Under that it was proposed that if an American publisher had a book that he wanted produced in Australia, or that his author insisted must be produced in Australia, he should stipulate in offering the book to a British publisher that the latter should either produce the book in Australia himself or do his best to place it with a local house; and that in either event the British publisher should pay the American publisher 70% of the Australian royalties.[1] The delegation sought also to achieve some uniformity in the open markets and proposed that neither country should, without the consent of the other, supply any book, published on both sides of the Atlantic, at less than its usual trade terms or sub-lease the rights of

[1] This plan, created by the war, ended with the war.

publication to a third party. But not much progress was made, and when the three sailed for home on 31 May their proposals had not got beyond the Directors of the Bureau and the Bureau seemed less able to commit its members than was the Association.

In Canada. In Canada the three delegates found that the attractive power of American civilization and education had greatly increased, and with it a strong preference for American format, particularly as compared with British war-time production. An open market in Canada commonly meant a market closed to the British edition, and they found also that some British publishers temporarily unable to supply their editions of books in which they held an exclusive Canadian market were permitting the American editions to enter the market for the duration of the war. There was a disastrous famine of British books. Even though the delegates were convinced that their colleagues at home must put Canada at the head of the ration queue, it would not be sufficient, and it quickly became their principal task to investigate the possibilities of local manufacture not only for sale in Canada but for export to South America and, whenever possible, to the United States with the assistance of a co-operative agency and distributing centre in New York. Man-power rather than paper was the immediate problem and some encouragement was received from the Labour Department in Ottawa upon which Canadian publishers might build.

The mission's report. The report which the three delegates presented[1] on their return received some interest outside the trade. Part odyssey, part document, it had literary quality and foresight and pays re-reading a quarter of a century later. Although the trade context in which it was written has inevitably changed it gives a picture of the war effort in the United States and Canada, of which the three were permitted by the sponsoring Government Departments to see as much as security allowed. In Canada they visited the R.C.A.F. Ferry Command aerodrome and the Vickers works near Montreal and the Uplands training station; in America they saw the Kaiser shipyard, where prefabricated Liberty ships were being built in twenty-eight days, and the San Diego naval station; they spent a day on one of the Moran tugs marshalling a convoy in New York harbour. For light relief they had a morning in Hollywood with Walt

[1] *Report by the three delegates of the Publishers Association sent to North America at the invitation of the Book Publishers Bureau and the Book Section of the Canadian Board of Trade.* London: The Publishers Association. Printed for private circulation 1943.

Disney and Donald Duck; and at a lunch given for them by the Writers War Board a feature was the reading by Clifton Fadiman of a poem which Ogden Nash had composed for the occasion.

The acceptance of the report by the Council in September 1943 was followed immediately by the formation of two committees, a Canada Committee[1] and an Anglo-American Relations Committee,[2] to pursue the possibilities which had been opened up.

The Canada Committee. The Canada Committee took up the possibilities of increasing book-production in Canada and had exploratory discussions with the High Commissioner in London, with the Premier of Ontario during a visit, and with the Dominions Office. But most of all its concern was to open the eyes of British publishers to the urgent need almost to start all over again in the Canadian market: to compete with American production and packaging, to give better sales promotion, to refuse to sell Canadian rights to American publishers, and to take action to keep out American book-club editions of books by British authors. To bring the situation home it was necessary to confront members of the Association with a representative delegation from Canada. But it became known that any delegation arriving early in 1944 would have to wait for many months before it could go home; and the landing in Normandy made the reason clear. It was, thus, September 1945 before the Canadian delegation arrived. Its leader was C. H. Dickinson (Ryerson Press) and he was accompanied by W. H. Clarke (Oxford University Press), F. D. Tolchard (Secretary of the Toronto Board of Trade) and C. R. Sanderson (Chief Librarian of the Toronto Public Libraries). From the visit members of the Association gained, at least, first-hand evidence of American standards of production and salesmanship, against which their own publications had to compete; and a hope that a co-operative distributing centre and showroom for British books might be established in Toronto to supplement the activities of the Canadian publisher-agents.

The U.S.A. Committee. The work of the U.S.A. Committee covered the whole field of Anglo-American competition in world markets. In the continuing negotiations with the American Bureau, of which it took

[1] Canada Committee: W. G. Taylor (Chairman), Walter Harrap, Geoffrey Faber, G. Wren Howard, G. F. J. Cumberlege, Lovat Dickson (Macmillan), H. Aubrey Gentry (Cassell), R. H. Code Holland.

[2] Anglo-American Relations Committee: B. W. Fagan (Chairman), Walter Harrap, Geoffrey Faber, G. Wren Howard, W. A. R. Collins, R. P. Hodder-Williams, Douglas Jerrold, Stanley Unwin.

charge, it emphasized that it had no concern with the many American books which failed to find a British publisher, but for those books published in both countries it sought to convince the Bureau that British publishers had a right, in principle, to retain the Empire markets (with the possible exception of Canada for books of American origin) and were better placed geographically to serve the European market. The justification of its position and suggestions for a common code of practice in open markets were set out in a series of Observations in April 1944.[1] The inevitability of American competition in export markets was also brought forcibly to the attention of the Board of Trade. When in May 1944 the Council submitted to the Board a report on the current state of the book-publishing trade in relation to current demands, it included as an appendix a long memorandum, prepared by the U.S.A. Committee, covering British-American publishing problems and their effect on the export trade in books. The help sought from the Government included the release of more paper to improve the quality and the appearance of books intended for export, greater freedom of imports (particularly of books produced in Canada) for re-export, and even some income-tax concession for American authors on royalties earned on sales in the United Kingdom. The U.S.A. Committee concerned itself also with the promotion of the sale in the United States of British books which failed to find an American publisher.

Agreement reached. Throughout 1944 it became increasingly clear that some understanding between the Association and the Bureau on the division of world markets must be found, and it was with some relief that the Council of the Association learned that Edward M. Crane (of the Van Nostrand Co.) was to visit this country in June 1945 as an official representative of the Bureau. It was fortunate also that the Association then had as its President B. W. Fagan, who had been chairman of the U.S.A. Committee. Edward Crane brought with him a draft of forms of publishing contract, to be jointly recommended by the Association and the Bureau, regulating the division of markets and the purchase of subsidiary rights. From the meetings which ensued there came an agreed code of practice which was approved by the Council of the Association and the Directors of the Bureau.[2] It provided, in particular, that the world markets would be divided into two categories: the 'home' market of the United States and British publishers respectively, in which each should have the exclusive right of sale, and 'other' markets, which should be open markets

[1] see *Members' Circular* xx, no. 4, 44–7.
[2] soon to become the American Book Publishers Council.

The export markets 197

to both; that the 'home' market should be: for the British publisher, the British Commonwealth, its colonies and dependencies; for the American publisher, the U.S.A., its colonies and dependencies; and that the 'other' markets should be the rest of the world. There might be two exceptions only, on the ground of proximity and efficient handling: the American publisher might have an exclusive market in Canada (and not only for books of American origin) and the British publisher might have an exclusive market in Europe when that seemed to be an advantage to the particular book. Although there was some compromise by the Council on Canada, in barter for Europe, it had got what it wanted—on paper. But not all American publishers were prepared to go along with the Bureau,[1] and 'the British Commonwealth, its colonies and dependencies' was not long to remain a static concept.

Export research. War-time contacts, for instance in Australia, and the unavailability of British books had given American publishers a new consciousness of the world's markets, and their Government, seeing books as instruments of the American way of life, gave them aid. Edward Crane came not only as a delegate of the Book Publishers Bureau, but as chairman of the United States International Book Association (USIBA), a new organization set up with some Government help to promote exports through a chain of agencies throughout the world. The most effective reply of British publishers was to set themselves to give a better service. To this end perhaps the most important decision of the Council of the Association in 1945 was to establish an Export Research and Development Service, for which the initiative came from the Secretary, Frank Sanders, and to appoint a committee to plan and guide it.[2]

TRADE RELATIONS AT HOME

War-time difficulties of travel between London and the provinces inevitably created problems of representation for the Associated Booksellers, but co-operation between the two Associations was not impaired. When in 1942 the A.B. proposed the formation of a single National Book Trade Association, to be endowed with the Book Tokens scheme,[3] the Council of the P.A. concluded that both associations must retain autonomous powers and that accordingly a new association could achieve little more than was being achieved by joint committees. But in spite of this rejection

---

[1] Indeed, the 'agreement' had to be formally disowned because of U.S. anti-trust legislation.
[2] R. J. L. Kingsford (chairman), R. H. Code Holland, G. Wren Howard, F. D. Sanders.
[3] see p. 144.

the work of the Joint Book Club Committee, the action taken in connection with the Co-operative Societies, and—paradoxically—the long-drawn-out and irreconcilable War Risks Insurance controversy strengthened the ties by extending the principle of co-operation which had been first effectively embodied in the now familiar Joint Advisory Committeee.

The Joint Advisory Committee. The principal task of the J.A.C., the consideration and listing of new booksellers, greatly increased as the war advanced and the ease of selling every kind of book became apparent. The number of applicants, which had been 304 in 1938, fell to 134 in 1940 and then grew to 573 in 1945 and leaped to a peak of 1,164 in 1946. In 1940 the Committee began to list its 'recognitions' in two classes: 'A' denoting businesses meriting publishers' direct attention, 'B' the smaller business which might be left to the wholesalers.[1]

As supplies of books became short the Council agreed during 1941 to recommend publishers to maintain a fair distribution to established customers, but to take note of war-time shifts of population, and in November of the same year further authorized the Committee to recommend the refusal of any applications for trade recognition if, after careful investigation, it was found that the locality was adequately served by existing booksellers. It was thought then that the work of the Committee might be affected by the Location of Retail Businesses Order 1941, which the Board of Trade introduced to meet shopping needs in newly populated areas. But the Board agreed to suggest to the Local Price Regulation Committees (who were made responsible for considering applications for licences under the Order) that they should co-operate with the J.A.C. when considering applications from those intending to operate bookshops and commercial circulating libraries; and by 1944 the Board appeared to accept the view that it was useless for the local committees to grant licences in the teeth of the J.A.C. Famine conditions also inevitably increased the theft of books from publishers' warehouses and bookshops and in 1945 by the publication of an article in the *Library Association Record* the Committee drew the attention of public librarians to the extent to which they were unwittingly making use of this source of supply.

Commercial lending libraries. The limited, or so-called 'other trader', recognition accorded to commercial lending libraries was the subject of some disagreement between the publisher and the bookseller members of the J.A.C. in 1945 and the Council of the Publishers Association upheld

---
[1] The classification and listing of wholesalers themselves was not initiated until after the war.

the view against the Associated Booksellers that 'full recognition' was appropriate to businesses whose sole activity was the sale or lending of books, even though the business might be confined to certain kinds of books only. The reluctance of the Associated Booksellers to accept booklending as a first step to bookselling had led in 1944 to the formation of a Commercial Lending Libraries Association, which sought representation on the J.A.C. and supported a proposal of the Fiction Group that as the sale of books was regulated by the Net Book Agreement, so lending should be regulated by a Net Lending Agreement. But the commercial lending library, other than the subscription library, was soon to disappear in the post-war world.

Partial remaindering. Early in 1939 the thorny subject of 'partial remaindering'[1] came under discussion at a meeting of The Publishers' Circle and following a subsequent recommendation from the Fiction Group, in March the Council appointed a committee under Sir Humphrey Milford to investigate the practice and its effects. He was regarded as the ideal chairman for an investigation of a practice of which the then President of the Association, Faber, was an advocate and the Vice-President, Wren Howard, was an outspoken opponent. But the Committee had held only one meeting when war broke out and the evacuation of part of the London business of the Oxford University Press took Milford to Oxford. Nevertheless it was thought that while war conditions put the practice into abeyance a solution might be reached, and in 1943 a committee[2] made a fresh start. In 1945 it proposed to the Council that those publishers who were in favour of 'partial remaindering' or had had recourse to it in the past should be invited to a meeting to consider the arguments that had been brought forward against its continuance and that those arguments should be circulated in a memorandum to be prepared by Wren Howard and Kingsford. The meeting was held on 31 January 1946 and the memorandum carried the day. Since it throws some light on Public Library purchasing at the time, two extracts from it may be of interest:

The Partial Remaindering Committee has been told beyond doubt by those library suppliers from whom it has taken evidence that the success of a partial remainder depends upon the retention of the original price on the jacket. The Public Librarian is, therefore, offered a spurious bargain. We recognise that some librarians will always endeavour to buy some proportion of their purchases at cut prices, e.g. review copies, genuine remainders, ex-library and second-hand

[1] see p. 149.
[2] Sir Humphrey Milford (Chairman), R. J. L. Kingsford (Deputy Chairman), G. C. Faber, C. W. Chamberlain, G. Wren Howard, Douglas Jerrold.

copies, but we do not accept the argument that it is useless to stop Partial Remaindering unless these other sources of supply are stopped also. Public Librarians cannot live on the limited supply of review copies, on second-hand copies and on publishers' failures and we see no reason whatever to supplement these sources by a peculiarly insidious bargain, the Partial Remainder, which more than the genuine remainder or the ex-library copy enables the Public Librarian to present himself to his Committee as an astute buyer of recent books at an abnormally high discount which is available to the privileged few and not the bookselling trade in general.

The advantage to be gained from Partial Remaindering has lain in its very dubious legitimacy...but if it is now to be pronounced legitimate, many of those whose library sales have suffered by their abstention (and it is not argued by library suppliers that Partial Remaindering has increased the total volume of library purchases) must inevitably follow suit sooner or later. No advantage will then remain with anybody—except the Public Librarian, who will have an ever increasing range of partial remainders to choose from until in effect the library discount will have been increased from 10% to the equivalent of 60% or even 70%.[1]

The meeting passed a resolution condemning 'partial remaindering' (defined as 'the practice of selling to selected booksellers net books, both fiction and non-fiction, for sale to the general public and/or Public Libraries at prices less than the net published price, while maintaining the net published price in all other directions') as undesirable in the interests of the trade and contrary to the spirit of the Net Book Agreement and asking the Council to outlaw the practice by the addition of a new clause to the Decisions and Interpretations relating to the Net Book Agreement.[2] The new clause was duly enacted.

National Book League. Much thought has been given since the end of the war, and continues to be given, to problems of distribution within the book trade and to the need, in the interest of both speed and economy, for clearing houses between bookseller and publisher through which orders could be sent, books could be dispatched and accounts could be paid. It is therefore of historical interest to note that in 1940 the National Book Council put forward to the Association a scheme for a clearing house for orders—a proposal which the Council regretfully judged to be impracticable as the bombing of London was intensified during that autumn and decentralization of every kind became an increasing need. The war, however, was not to delay the evolution of the National Book Council itself. In February 1944 Geoffrey Faber, one of the Association's repre-

[1] 'Against Partial Remaindering', P.A. memorandum 1945.
[2] This decision was confirmed by the Trade Descriptions Act, 1968.

sentatives on the Executive Committee of the N.B.C., informed the Council of plans to turn the N.B.C., which was then mainly the public relations instrument of the trade, with the Publishers Association as its principal godfather and source of income, into the National Book League as we now know it. The scheme went forward and in December 1944 John Hadfield, the first Director of the N.B.L., addressed the Council of the Association upon the policy of the League. Although the Association was no longer officially represented in the new constitution, the League was to draw many chairmen from the ranks, and much of its income from the pockets, of its publisher members.

Society of Authors. War-time conditions inevitably exerted some strains in the relations of the Association and the Society of Authors, and in one matter of controversy heat was generated by the insensitiveness of the Council of the Association to the claims of the Society to represent authors other than its own members and by the epistolary style favoured by the Secretary of the Society, D. Kilham Roberts.[1] In January 1942 it had come to the Council's notice that some authors, or literary agents acting on their behalf, were claiming under the 'reversion of rights' clause in their publishing agreements that their publisher should surrender his rights in books which were temporarily out of print, even though the publisher was prevented from reprinting by circumstances beyond his control. 'Many publishers' the Council stated in a memorandum[2]

would find themselves in an exceedingly awkward position if authors were to insist on the strict letter of their contracts, and invariably to demand cancellation of such contracts, if the publisher found himself unable to reprint a book within the time limit fixed. A publisher who has had considerable stocks destroyed by enemy action cannot hope to have enough paper to reprint all his destroyed stocks of books within any specified limit of time. Indeed, it is nowadays often impossible for a publisher to keep anything like the whole of his catalogue in print, even though he has suffered no such loss... The Council is of the opinion that it would be grossly unfair for authors, or their agents, to take advantage of the present difficult conditions by the unreasonable exercise of their legal rights to secure the cancellation of publishing agreements, the provisions of which may include valuable options on future work. On the other hand, authors cannot reasonably be expected to wait indefinitely—that is to say, until some date after the conclusion of the war—for their books to be put back into print if there is any appreciable demand for them.

[1] The Secretary of the Society to the President of the Association, 24 April 1944: 'The facts speak for themselves and you won't change them by writing me silly and provocative letters.'
[2] *Members' Circular* XVIII, no. 1, 9.

To ensure therefore that one publisher did not take advantage of another's misfortunes and that authors' legitimate interests should be protected, the Council proposed that members should bind themselves not to take over any book from another member until all the circumstances had been investigated by an advisory tribunal to be appointed by the Council. Within a month such an agreement had been signed by 121 of the 150 members. The rights of authors under the out-of-print clause in their contracts were not touched by the agreement, but there was a potential loss of freedom in bargaining, and it was the Officers of the Association who were designated to form the tribunal. In July 1942 although no case had yet arisen for consideration during the five months of the agreement's existence, the Society of Authors objected to it and in September a meeting was held at which it was agreed that the Officers of the Association would consult the Society before making a decision if the author concerned was a member of the Society and would inform the Society of the decision made when it concerned an English author who was not a member of the Society. The Association undertook also to redraft the agreement so as to limit its existence to the duration of hostilities with Germany and six months thereafter; and finally it was agreed that if the Association felt the necessity for any other agreement concerning authors' contracts it would consult the Society 'in a friendly spirit of co-operation'. But the friendly spirit was not promoted by correspondence and it was January 1945 before the Officers again met representatives of the Society's Committee of Management. There was a general discussion of coming opportunities in overseas markets and of the retention by British publishers of their rights in these markets, and the Society offered some hope that it would support the publication by British publishers of cheap continental editions.[1] The meeting failed to reach agreement upon a scale of fees to be charged for the use of poems and extracts in anthologies; and the Society was unwilling to join with the Association in proposing to the Board of Trade that for books first published during the war the Copyright Act of 1911, then in force, should be modified to increase the period of unrestricted copyright from twenty-five to thirty years and to decrease the following restricted period from twenty-five to twenty years.[2]

[1] In August 1944 the Books and Periodicals Commission of the Allied Ministers of Education (see p. 182) had given warm encouragement to the possibility of a British counterpart to the pre-war Tauchnitz and Albatross series.

[2] see p. 41. The restricted period was abolished by the 1956 Copyright Act, which provides an unrestricted term of fifty years after the author's death.

The coming of the microfilm. For the duplication of records against the risks of bombing and for the reproduction of the ever-increasing number of out-of-print books microphotography provided an answer; and in October 1942 a talk which the Treasurer of the Association had in Cambridge with Professor R. S. Hutton, an active member of the Association of Special Libraries and Informational Bureaux (ASLIB), alerted the Council to the importance of regulating the making of microfilms in relation to copyright in literary material. Early in 1943 a meeting took place between representatives of the Association, the Society of Authors and ASLIB and a memorandum of agreement subsequently prepared by Wren Howard, the Association's representative, was adopted by all three bodies. In it they agreed that all films which consisted of reproductions of copyright books or pamphlets should be manufactured to the order of the publisher of the work, who would make the sale and arrange for the payment of an appropriate fee to the copyright owner. The authorized films would be manufactured by ASLIB, which would also act as a bureau to receive all applications from its members and other bodies and pass them on to the publisher. Every microfilm of a copyright work should include the title-page, with the name of the author and publisher, and at intervals throughout its length a notice 'that this microfilm is copyright and must not either be copied or be used for public exhibition in any manner, either in whole or in part'. Nevertheless, the degree of copying by photographic or other means permitted by the 1911 Copyright Act remained in doubt and in 1944 the medical publishers were expressing some alarm that the documentation service offered by the Royal Society of Medicine to its members included not only information about material available in its library but the supply of copies of it.

International copyright. In the wider field of international copyright, it was a matter of some irony that of the four allied nations the United States, the Soviet Union, and China remained outside the international convention. Between Britain and the United States there was a temporary reciprocal understanding, covering books first published during the war. In March 1944 by a British Order in Council the period of fourteen days allowed under the 1911 Copyright Act for the 'simultaneous' publication necessary to secure the protection of that Act was extended to the period of the war and one year thereafter, in respect of works first published in the United States during that period; and in the United States a Presidential Proclamation extended the benefits of the U.S. Copyright Act to works of

British nationals first produced outside the United States during the war. A Chinese goodwill mission to Britain in January 1944 included the Chairman of the Chinese Publishers Association, Mr Wang Yun-wu, who was entertained to tea during a Council meeting and was optimistic about the possibility of China joining the Copyright Union after the war, provided that special dispensation could be given for translations into Chinese. A new Chinese copyright law duly followed later in the year, but although it ostensibly offered copyright protection to all publications, whether in Chinese or any foreign language, the Board of Trade advised the Association that the Chinese Government was still not prepared to suppress pirated translations.

Printers and binders. The establishment in April 1941 by the Ministry of Labour of the Book-Production Advisory Committee,[1] drawing three of its members from the Association and two from the British Federation of Master Printers, provided a welcome opportunity for the two bodies to work in unison; and in September 1941 this example was followed in the appointment of a standing Joint Consultative Committee composed of the three officers of the Association and representatives of the book-printing and bookbinding sections of the Federation. Both inside and outside the committee room it developed a common policy towards the emergencies of the war—scarcity of labour, the supply of materials, the collection of metal for salvage. The more permanent problems of printing, binding, and warehousing charges caused some disagreement, and on occasions it seemed to the Officers of the Association that it was the policy of the Federation to announce higher charges first and then, after discussion at a meeting of the joint committee, to expect the Association to recommend a *fait accompli* to its members.

Binding research. The Association and the Federation found common ground in a new enterprise. In September 1944 the President of the Association advocated the coming importance of the trade associations initiating technical research, for which Government grants could be obtained, and he put forward a project—dear to the hearts of himself and Wren Howard—for a programme of research to be undertaken by the Printing and Allied Trades Research Association, in the methods of bookbinding, which were basically not dissimilar from those familiar to Caxton, and in the development of new materials such as plastics. The advisory and financial partnership of the Master Binders section of the

[1] see p. 171.

Federation was readily secured and PATRA[1] was launched into a programme of research on the strength and warping of covers and their vulnerability to attack by mould and insects, on lettering processes, and on the development of new materials and adhesives. Moulds were duly grown and insects were enlisted, but little more was achieved.

Domestic affairs. It has been seen[2] that when Export Groups in all branches of industry were formed in 1940, membership of the Export Group of the book trade became a condition for receiving the full paper ration. Some members of the Group, seeing what they owed in this respect to the work of the Association and anticipating other help from it, became members of the Association itself. Although the war forced some members out of business, the membership rose from 124 in 1939 to 202 in 1945. The entry by Walter Hutchinson of the eleven firms in his group in 1939 and his withdrawal of them in March 1942, as a protest against the Association's participation in the operation of the Moberly Pool, caused some temporary fluctuation. At the beginning of 1943 a new basis for members' annual subscriptions was introduced, consisting of a basic subscription with an additional rate for each employee in excess of ten but subject to a maximum of 100. Separate subscriptions to the Groups within the Association were abolished, though an appeal for funds might be made by a Group to its members.

The activities of the Groups inevitably diminished as many of their problems, arising from war conditions, fell directly on the shoulders of the Council. The Fiction Group, the only one having to work within fixed prices, tried hard in 1940 to secure a uniform policy in advancing the prices of 7*s.* 6*d.*, 8*s.* and 8*s.* 6*d.* novels to meet the cost of war risks insurance and rising costs of manufacture. The Medical Group was active, sometimes to the embarrassment of the Council, in advocating the superior claims of medical books for more paper, and the Educational Group similarly for school books. In the distinguished line of chairmen of the latter Group H. G. Wood (of Nisbet's) was succeeded in 1945 by A. W. Ready (of George Bell & Sons). Rises in the cost of living and the scarcity of labour required the attention of the Employment Group and its Wages and Disputes Committee and in 1941 the agreed minimum wages of warehouse employees began to be supplemented by cost-of-living bonuses. One new Group was formed, for Religious Books, in 1944.

[1] In 1967 PATRA amalgamated with the research organization of the paper and board industries to become The Research Association for the Paper and Board, Printing and Packaging Industries, generally known as PIRA.   [2] see p. 166.

The Association's jubilee. In 1945 the President called the attention of the Council to the coming of the Association's fiftieth anniversary in the following year and it was agreed that the event should be commemorated by the publication of a historical record. But desirable authors had more pressing calls on their time and it remained for the President of 1963–5, John Brown, Publisher to the University of Oxford, *faute de mieux* to invite that President of twenty years earlier to attempt the task himself.

Epilogue
The 1962 verdict

> It seems to us certain that if the Net Book Agreement ceased to operate fewer books would be published...inevitably, we think the effects would be more severe in the higher reaches of literature...Net prices are fixed by publishers in conditions of free competition. The main object of the agreement is, in our view, to preserve retail price stability: it is not an instrument for fixing prices.
>
> Mr Justice Buckley, 1962

Although the production of paper was severely curtailed by the shortage of coal and the bitter weather during the winter of 1946–7, rationing ended in March 1949. It was an apparent symbol of the ending of governmental restrictions, and publishers found themselves in possession of an independence of action which many had almost forgotten and some had never known. Towards the end of 1950 the superior currency appeal of the dollar, which deprived British paper-makers of much of their raw materials, the Government's priorities for its new defence programme, and a deficiency of home-produced straw after a poor harvest brought a fresh shortage of paper and boards, which was the subject of a debate in the House of Commons. Nevertheless a reintroduction of control was avoided. With freedom came new problems: of rising costs, as the scarcity of skilled operatives in printing brought higher wages and as the cost of materials increased after the devaluation of sterling in 1949, and of diminishing sales of individual titles. Scarcity had increased the appetite and almost any book for which a publisher could find the paper had been oversubscribed before publication. Now books had again to be *sold*.

The course of exports did not run smoothly. European sales were limited by currency restrictions and there was a deterioration in the relations between British and American publishers on the division of world markets for books published in both countries. USIBA was short-lived, but the joint statement on markets agreed between the Association and the American Book Publishers Bureau achieved little. British publishers were determined to hold on to the traditional markets which they had cultivated and which were economically vital to them, whether these markets were a part of the current or the historical British Empire. American publishers were inclined to equate the British market with the disintegrating Empire and were moreover irritated that the need to save dollars had led the British Government to control the importation of American fiction. In Canada the British position was barely held. The proposal for a wholesale organization for British books in Canada had to be abandoned; of those Canadian publishers who acted as agent for both British and American publishers some saw more butter on the American side of the bread and were fearful of being thought too pro-British.

In spite of the difficulties the proportion of export sales to total sales grew within the decade 1946–56 from 25 to 37%. But few books are published for export only. Vital as is the export trade, no less to the publishing economy than to the national, it is upon initial success in the home market that the export sale of most books depends.[1] By 1950 there was a perceptible fall in the sales of individual titles and the beginning of a steep rise in the costs of production, and the one reacted on the other. Book prices did not rise comparatively with the prices of other commodities, but every thinking publisher was brought up abruptly against the problem of the widening disparity between the replacement costs and the historical costs of his stock. With active memories of war-time price control and with the disincentive of high taxation, most were disinclined to risk charges of profiteering by taking in more from today's sales to meet tomorrow's higher replacement costs, and for private companies the problems of finding more and yet more capital to finance growth and inflated costs in the coming years were consequently to be the greater. In the shops there was a visible disparity between the prices of the new publication and the 'back list' and booksellers, desperately anxious to retain their war-time high level of turnover and yet seemingly convinced that it was impossible to do so, sought remedies in improved conditions and terms of supply and in the restriction of bookselling outlets.

It was against this background that a new joint Book Trade Committee, on the lines of that of 1927, worked during the years 1948–51. The problems of that Committee and the submission of evidence in 1948 to the Board of Trade's committee of inquiry into resale price maintenance required the Association to rethink its attitude to the other half of the trade. There emerged a new faith in bookshops as the best and the most economical means of selling books, and after some equivocation a deter-

---

[1] School books for developing countries, calling for special publishing expertise, are an obvious exception.

mination that well-informed stockholding booksellers could not exist without the protection of the Net Book Agreement.

The complete record of the defence of the Net Book Agreement before the Restrictive Practices Court in 1962—the preparations, the documents, the hearing, the judgment—is available elsewhere[1] and does not need retelling in detail here. Although publishers might act through the years with some uniformity in fixing their prices for certain kinds of books, in particular the novel, the Net Book Agreement was concerned not with price fixing, but solely with price maintenance. Even so, after 1948 the Association became aware that some of the machinery developed for its enforcement—the listing of 'recognized' booksellers and the stop-listing of infringers—was contrary to current public opinion and before the Restrictive Trade Practices Act became law in October 1956, steps had been taken to adjust it. The Act itself prohibited, in particular, the enforcement of prices by collective action; and the Agreement was amended accordingly. The new Net Book Agreement was approved at a Special General Meeting of the Association on 18 February 1957. It was a fact that no bookseller had been black-listed for years; no doubt the threat of stop-listing had proved itself a sufficient discipline. In its place the new Agreement rested upon publishers' standard conditions of sale of net books and upon their appointment individually of the Association as their legal agent empowered to call upon any one whose conditions of sale had been breached to take action in defence of those conditions, such action to include an application to the Court for an injunction to restrain the infringing bookseller. The Restrictive Trade Practices Act even legalized the enforcement of conditions of sale through a wholesaler to a retailer and in this way strengthened the net book system. The other great change was in the approval of booksellers. Since the system of recognition, on the recommendation of the Joint Advisory Committee, had been introduced in 1929, its limitation had never been absolute; publishers had individually refused to supply some booksellers who were on the list and had supplied others who were not. Nevertheless in 1957 any suggestion of recognition or compulsion was removed and the list of recognized booksellers was superseded by a *Directory* containing, as a service to publishers, the names and addresses of those firms which the Joint Advisory Committee considered to be *bona fide* booksellers. At the

---

[1] *Books are Different*: an account of the defence of the Net Book Agreement before the Restrictive Practices Court in 1962. Edited by R. E. Barker and G. R. Davies (Macmillan 1962). A brief account is also given in J. J. Barnes: *Free Trade in Books* (Oxford: at the Clarendon Press, 1964), pp. 153-72.

same time agreements within the Fiction and Map Groups limiting trade discounts were abandoned. Counsel for the Registrar was to say with some tartness in his cross-examination in 1962: 'and everything went on much the same.' Indeed, the old Net Book Agreement was so little restrictive that not much change was necessary to make it defensible under the Act.

The Association had to ask itself not only whether the Net Book Agreement was defensible, but whether it was worth defending. The instinctive affirmative (no living publisher had worked under any other system) had to be rationalized against the certainty that the cost of defence would be very great in time and money[1] and the possibility that even if the defence were successful, the Government might introduce further legislation to make resale price maintenance itself illegal. Was the general support of the net book system by members of the Association sufficiently critical and determined? What would be the attitude of the public to the defence and, if the defence failed, what would be the effect on the prestige of the Association? These heart-searchings caused some disturbance of feelings between the Association and the Booksellers Association, who after a pre-1956 refusal to believe that the Net Book Agreement could be in jeopardy (had not books been the only commodity to escape the imposition of purchase tax in 1940?) were subsequently inclined to accuse the Association of being insufficiently concerned with its fate. But these differences were in the end the means of bringing the two Associations closer together, and from the trial was to come a combined determination to raise the quality of bookselling.

Although the Net Book Agreement, including the Public Libraries and Book Agents schemes, was registered in 1957, it was not until July 1959 that the Registrar of Restrictive Trading Agreements informed the Association that he was about to refer the Agreement to the Court and the Council had to decide whether to take up the gage. The omens looked bad; only one out of seven agreements—and that an unusual export one—had survived the test and several hundred had been abandoned. 'It was not mere conservatism' wrote the President, R. W. David,[2] in the 1959–60 *Report of the Council*,

that induced the Council to decide, unanimously, that the Agreement must be defended at all costs. The Council believes that however strong may be the arguments for free trading in general, they are not applicable under the unique conditions of the book trade. The wares in which publishers and booksellers are

[1] It was £36,440, of which the Booksellers Association contributed £15,176.
[2] Cambridge University Press.

dealing exist in 300,000 different varieties, each of considerable complexity in itself, and each year brings 20,000 new specifications to be mastered. To bring each book to the purchaser who really needs it, to find for each purchaser the book he needs, requires an elaboration of 'services' (extensive stock, technical knowledge, the ability to research and to obtain to order) that no retailer can readily undertake unless his margins are reasonably assured beforehand. If the bookseller were not prepared to maintain this necessarily complex machinery for distribution, the course of all but the few 'best-sellers' from publisher to public would be seriously impeded and both would suffer.

At a special General Meeting in September 1959 the members of the Association decided, without a dissentient voice, to defend the Agreement; and a Defence Committee[1] was formed to prepare the case. The Proof of Evidence which it produced—largely the work of R. E. Barker—was to run to some 70,000 words.

The Association was not technically committed to the defence until its Annual General Meeting in March 1962, when the date of the hearing had just been announced. In the intervening $2\frac{1}{2}$ years most associations appearing before the Court had lost their cases, the Association had lost its two leading Counsel upon their elevation to the Bench, and the talk of Government action to end all resale price maintenance had increased. Nevertheless the members confirmed their decision to face the Court; and into Court on 25 June 1962 the Association went, represented by Arthur Bagnall, Q.C., D. A. Grant, Q.C., and Jeremy Lever, to be heard before Mr Justice Buckley, Sir Stanford Cooper, W. L. Heywood and D. A. House, with H. A. P. Fisher, Q.C., and R. H. W. Dunn appearing for the Registrar.

The hearing lasted until 27 July and on 30 October Mr Justice Buckley delivered the Court's judgment. It was a total victory for the Association, not only in its concluding declaration that the Net Book Agreement was not contrary to the public interest, but in its justificatory analysis of the nature of publishing and bookselling. The basis of the judgment was that the production and marketing of books involve problems that are different from those which arise with other commodities; and indeed no other case in which resale price maintenance was a central issue had succeeded before the Court. The verdict of Lord Justice Campbell in 1852 had been reversed.

The effect of the judgment on the book trade was far-reaching, in self-criticism and in determination to extend the full service of booksellers.

---

[1] The Committee during the intensive preparation and the defence itself was: The President, J. T. Boon (Mills and Boon); the Vice-President, R. W. David; the Treasurer, John Brown; R. H. Code Holland; the past and present Secretaries, F. D. Sanders and R. E. Barker. Barker had succeeded Sanders as Secretary in 1958.

'A number of people' wrote J. T. Boon, the Association's indefatigable G.O.C. during the action, 'had been forced for a period of years closely to scrutinize the trade and the way in which it functioned, not in isolation but in relation to the public good and to other industries. From this sprang a much more self-conscious, healthily self-critical approach.'[1] One immediate consequence was a seminar in which the economists who had given evidence for the Association, P. W. S. Andrews, and his assistant Elizabeth Brunner, explained in detail to leading publishers the characteristics of the trade as they appeared to economists and suggested ways in which booksellers could be encouraged and helped to carry out yet better those functions on which the service of the public and their own prosperity depended. Through the case the Publishers and Booksellers Associations had achieved a better *entente* than at any time since price-cutting and the accompanying decline of retail bookselling had caused Frederick Macmillan to fix a net price for Marshall's *Principles of Economics* and had led to the founding of the Publishers Association in 1896.

[1] *Books are Different*, p. 55.

Appendix 1
The first Rules of the Association

OBJECTS

The objects of the Association are:
  To promote and protect by all lawful means the interests of the Publishers of Great Britain and Ireland.

RULES

Persons eligible

I. All firms, companies, or individuals who for not less than one year have carried on the work of *bona fide* book publication shall be eligible for membership. Those whose business consists mainly of the sale of books at retail, or who belong to any association or society of retail booksellers, shall not be eligible. The publication of newspapers or periodicals shall not be regarded as a qualification.

Election of members

II. Application for membership shall be made to the Secretary, who shall refer such application to the Council. The decision of the Council as to the eligibility of the applicant shall be final.

Subscription and entrance fee

III. An entrance fee of Five Guineas shall be payable by each firm or company on joining the Association, and a subscription of Five Guineas per annum payable in advance on or before the 1st of January in each year. Any partner, director, manager, or secretary of a firm or company, may attend and speak at General Meetings, and be eligible as an officer or member of the Council, but only one representative of each firm or company shall vote or be eligible.

Officers of the Association

IV. The officers of the Association shall consist of a President, a Vice-President, and an Honorary Treasurer, who shall be elected annually by ballot at the General Meeting, and shall be eligible for re-election, but the President and Vice-President shall not serve for more than two years consecutively.
  The officers shall be *ex officio* members of the Council.

The Council

V. At the Annual General Meeting a Council of ten members, in addition to the three officers, shall be elected by ballot for the ensuing year, and be eligible for re-election. No firm shall be represented on the Council by more than one member.

Five members shall form a *quorum*.

In case of necessity, the Council shall have power to fill any vacancy among the officers, or in the Council, until the next General Meeting.

Duties of the Hon. Treasurer

VI. The duties of the Honorary Treasurer shall be to receive and pay into a Bank to the credit of the Association all moneys payable to the Association, and to make any disbursements sanctioned by the Council.

All cheques shall be signed by the Honorary Treasurer, and countersigned by another member of the Council.

Duties of the Secretary

VII. The duties of the Secretary shall be to collect all subscriptions, donations, or other moneys due to the Association, and to pay the same over to the Honorary Treasurer. To organise the Association, and to endeavour to obtain the support and adherence of Publishers, and otherwise promote the objects of the Association, subject to the directions of the Council. To conduct generally all the correspondence and business of the Association, to keep proper books of accounts, and make minutes of the proceedings of all meetings of the Association or its Committees and to prepare and submit to the Auditors annually the accounts and vouchers, with a balance-sheet containing an accurate statement of the finances of the Association.

Auditors

VIII. Two Auditors shall be elected annually at the General Meeting, whose duty it shall be to make an audit of the accounts and balance-sheets of the Association, which, with proper vouchers, shall be submitted to them not less than fourteen days before the Annual General Meeting.

Annual General Meeting

IX. An Annual General Meeting of the members shall be held at some convenient place to be appointed by the Council, at which a report of the proceedings of the Association and the accounts and balance-sheet for the year shall be submitted for approval and adoption.

Ten days' notice of the Annual General Meeting shall be given.

Nine shall form a quorum.

Meetings of the Council

X. The Council shall meet quarterly, and special meetings may be summoned at any time by a notice signed by the President or Vice-President, on a requisition signed by two members of the Council.

Special General Meeting

XI. On receiving a requisition signed by not less than seven members of the Association, the Council shall within fourteen days convene a Special General Meeting, to be held at such place as they shall fix. The notice of such meeting shall state the nature of the business to be considered and the resolutions to be proposed, and no other business shall be entertained.

The Council may at any time direct a Special General Meeting to be called.

Arbitration

XII. Any member may request the Council to intervene in any dispute between such member and any other person or persons on a matter connected with his business, and act as arbitrators or appoint an arbitrator.

In any such cases of intervention, the member asking the assistance of the Council shall pay all legal and other expenses incurred by the Council.

Expulsion of Members

XIII. The Council shall have power to remove from the list of Members the name of any firm, company, or individual who may become bankrupt or in their opinion act in any way detrimental to the interests of the Association, or whose subscription shall be more than one year in arrear, provided two-thirds of the Council are present when the vote is taken.

Any firm, company, or individual so removed from the list of Members of the Association shall have power to appeal against the decision of the Council at the next General Meeting, provided that notice of intention to appeal be given within three weeks of the decision of the Council being communicated.

Alteration of Rules

XIV. No alteration in, or addition to, these Rules shall be made, except at the Annual General Meeting, or at a Special General Meeting convened for that purpose.

Three weeks' notice of such proposed alteration to be given to the Secretary in writing.

XV. The decision of the Council shall, in all matters not definitely provided for by the foregoing Rules, be final and conclusive.

Appendix 2
The members in 1896 and the first Council

Allen, George
Arnold, Edward
Arrowsmith, J. W.
Bell, George, & Sons
Bentley, Richard, & Son
Black, Adam & Charles
Blackie & Son, Ltd
Blackwood, William, & Sons
Bliss, Sands & Co.
Burns & Oates, Ltd
Cambridge University Press
Cassell & Co. Ltd
Chapman & Hall, Ltd
Chatto & Windus
Clark, T. & T.
Clive, W. B.
Constable, Archibald, & Co.
Dean & Son, Ltd
Dent, J. M., & Co.
Digby, Long & Co.
Gardner (Wells), Darton & Co.
Gay & Bird
Hachette & Co.
Heinemann, William
Henry, H., & Co. Ltd
Hodder & Stoughton
Houlston & Sons
Hurst & Blackett, Ltd
Hutchinson & Co.
Innes, A. D., & Co.
Isbister & Co., Ltd
Jarrold & Sons
Lane, John
Lawrence & Bullen
Lockwood (Crosby) & Son
Longmans, Green & Co.
Low (Sampson), Marston & Co., Ltd
Macmillan & Co., Ltd
Methuen & Co.
Murray, John
Nelson, T., & Sons
Nimmo, J. C.
Oliphant, Anderson & Ferrier
Osgood, McIlvaine & Co.
Partridge, S. W., & Co.
Paul (Kegan), Trench, Trübner & Co., Ltd
Prentland, Young, J.
Rivington, Percival & Co.
Routledge, George, & Sons, Ltd
Simpkin, Marshall, Hamilton, Kent & Co., Ltd
Skeffington & Son
Smith, Elder & Co.
Sonnenschein (Swan) & Co., Ltd
Stanford, Edward
Unwin, T. Fisher
Virtue, J. S., & Co., Ltd
Ward, Lock & Bowden, Ltd
Warne, F., & Co.

From 58 in 1896 the number increased to 68 in 1900, to 89 in 1910, to 93 in 1920, to 107 in 1930, to 126 in 1940, to 327 (272 full members + 55 associate members) in 1950, and to 363 (283 + 80) in 1960.

Appendix 3
The Officers and Secretaries of the Association, 1896–1962

Date	President	Vice-President	Treasurer
1896–8	C. J. Longman	John Murray	Fredk. Macmillan
1898–1900	John Murray	C. J. Longman	Fredk. Macmillan
1900–2	Fredk. Macmillan	John Murray	C. J. Longman
1902–4	C. J. Longman	Fredk. Macmillan	John Murray
1904–6	Reginald J. Smith, K.C.	William Heinemann	C. J. Longman
1906–9	Edward Bell	C. J. Longman	William Heinemann
1909–11	William Heinemann	Edward Bell	Arthur Waugh
1911–13	Sir Fredk. Macmillan	William Heinemann	James H. Blackwood
1913–15	James H. Blackwood	Sir Fredk. Macmillan	John Murray, C.V.O.
1915–17	Reginald J. Smith, K.C.	James H. Blackwood	W. M. Meredith
1917–19	W. M. Meredith	Humphrey Milford	G. S. Williams
1919–21	Humphrey Milford	W. M. Meredith	C. F. Clay
1921–3	G. S. Williams	Humphrey Milford	C. F. Clay
1923–4	C. F. Clay	{ G. S. Williams / Humphrey Milford	H. Scheurmier
1924–5	C. F. Clay	G. S. Williams	H. Scheurmier
1925–7	H. Scheurmier	G. S. Williams	G. C. Rivington
1927–9	{ W. M. Meredith / Edward Arnold	H. Scheurmier	G. C. Rivington
1929–31	W. Longman	G. C. Rivington	Bertram Christian
1931–3	Bertram Christian	W. Longman	Stanley Unwin
1933–5	Stanley Unwin	Bertram Christian	W. G. Taylor
1935–7	W. G. Taylor	Stanley Unwin	G. Wren Howard
1937–9	G. Wren Howard	W. G. Taylor	Geoffrey C. Faber
1939–41	Geoffrey C. Faber	G. Wren Howard	Walter G. Harrap
1941–3	Walter G. Harrap	Geoffrey C. Faber	R. J. L. Kingsford
1943–5	R. J. L. Kingsford	Walter G. Harrap	B. W. Fagan
1945–7	B. W. Fagan	R. J. L. Kingsford	R. H. C. Holland
1947–9	R. H. C. Holland	B. W. Fagan	J. D. Newth
1949–51	J. D. Newth	R. H. C. Holland	Kenneth B. Potter
1951–3	Kenneth B. Potter	J. D. Newth	Ralph Hodder-Williams
1953–5	Ralph Hodder-Williams	Kenneth B. Potter	J. Alan White
1955–7	J. Alan White	Ralph Hodder-Williams	Ian M. Parsons
1957–9	Ian M. Parsons	J. Alan White	R. W. David
1959–61	R. W. David	Ian M. Parsons	John Boon
1961–3	John Boon	R. W. David	John Brown

SECRETARIES

1896–1933	William Poulten
1934–58	F. D. Sanders
1958–	R. E. Barker

THE FIRST COUNCIL

PRESIDENT: C. J. Longman (*Longmans, Green & Co.*)
VICE-PRESIDENT: John Murray (*John Murray*)
TREASURER: Frederick Macmillan (*Macmillan & Co.*)
HONORARY SECRETARY: R. B. Marston (*Sampson Low, Marston & Co.*)
Reginald J. Smith, Q.C. (*Smith, Elder & Co.*)
Richard Bentley (*R. Bentley & Son*)
William Heinemann (*Heinemann*)
William Blackwood (*W. Blackwood & Sons*)
Edward Bell (*Geo. Bell & Sons*)
Slingsby Tanner (*A. D. Innes & Co.*)
John Hamer (*Cassell & Co.*)
T. Fisher Unwin (*Fisher Unwin*)
Oswald Crawfurd (*Chapman & Hall*)

Bibliography

UNPUBLISHED

The Publishers Association:
  Minutes of General Meetings, of the Council, of Groups, of committees.
  Annual Reports of the Council.
  Members' Circular.
  A memorandum on the Book Publishing Trade and the Schedule of Reserved Occupations, submitted to the Ministry of Labour and National Service, 12 February 1940 (ref. 17/40).
  A memorandum on the export trade in books, submitted to the Export Council, 5 April 1940 (ref. 49/40).
  A memorandum on the Purchase Tax and books, submitted to the Treasury, 15 May 1940 (ref. 87/40).
  A memorandum submitted to the Paper Controller, 24 May 1940 (ref. 97/40).
  A memorandum on War Risks Insurance of Publishers' stocks, submitted to the Board of Trade, 4 February 1941 (ref. 9/41).
  The Book Production War Economy Agreement, December 1941 (ref. 205/41).
  Memorandum on 'The British Book Industry', October 1941 (ref. 168/41).
  Report by the three delegates of the Publishers Association sent to North America at the invitation of the Book Publishers Bureau and the Book Section of the Canadian Board of Trade, 1943 (printed).
  Report on the current state of the Book Publishing Trade in relation to coming demands, May 1944 (ref. 91/44).
The London School of Economics and Political Science: British Library of Political and Economic Science (Pamphlet Collection).

PUBLISHED

A Report of the Proceedings of a Meeting (chiefly consisting of authors) held May 4th at the house of Mr John Chapman, 142 Strand, for the purpose of hastening the removal of the Trade Restrictions on the commerce of literature. (London: John Chapman, 142 Strand, MDCCCLII.)

R. Maclehose. *The Report of the Society of Authors on the Discount Question: a criticism.* (Glasgow: Maclehose and Sons, 1897.)

Publishers and the Public. (*The Times*, 1906.)

'*The Times*' *and the Publishers.* (The Publishers Association, 1906.)

Sir Frederick Macmillan Kt. *The Net Book Agreement 1899 and The Book War 1906–1908.* Two chapters in the History of the Book Trade, including a narrative of the dispute between *The Times* Book Club and The Publishers Association by Edward Bell, M.A., President of the Association 1906–8. (Glasgow: printed for the author, 1924.)

The History of 'The Times'. Vol. III, *The Twentieth Century Test, 1884–1912*. (London: *The Times*, 1947.)

F. D. Sanders, ed. *British Book Trade Organization: a report of the work of The Joint 1926–28 Committee*. (London: Allen and Unwin, 1939.)

James J. Barnes. *Free Trade in Books: a Study of the London Book Trade since 1800*. (Oxford: at the Clarendon Press, 1964.)

W. G. Corp. *Fifty Years: a Brief Account of the Associated Booksellers of Great Britain and Ireland, 1895–1945*. (Oxford: Blackwell, n.d.)

F. A. Mumby. *Publishing and Bookselling: a History from the Earliest Times to the Present Day*. (London: Jonathan Cape. Revised ed. 1949.)

Marjorie Plant. *The English Book Trade: an Economic History of the Making and Sale of Books*. (London: Allen and Unwin, 1939.)

Sir Stanley Unwin. *The Truth about a Publisher*. (London: Allen and Unwin, 1960.)

R. E. Barker and G. R. Davies, eds. *Books are different: an account of the defence of the Net Book Agreement before the Restrictive Practices Court in 1962*. (London: Macmillan, 1966.)

The Bookseller (title 1928–33: *The Publisher and Bookseller*).

The Publishers' Circular (to 1959).

Index

Acland, A. D. (W. H. Smith & Sons), 45
Agreements
  model forms of, 11, 12, 71
  Guide to Royalty, 126
  terminable, 127
Air raids
  destruction, 167, 170, 180, 187
  precautions, 159
Alden, H. E., 99 n., 142
Allen & Unwin Ltd, 54 n., 160 (and see Unwin, Sir Stanley)
Anderson, G. O. (Harrap), 37 n., 126 n., 162 n.
Anderson, Sir Hugh, 64–5
Andrews, P. W. S., 212
Arnold, Edward, 57 n., 60, 86, 86 n., 103 n.
ASLIB, 203
Athenaeum, The, 14
Austen-Leigh, R. A., 68
Australia
  booksellers, 109–11
  Broadcasting Commission, 125
  duties and import control, 112, 115, 191, 192
  economic difficulties, 11, 112
  Education Boards, 156
  market rights, 11, 118, 121, 191–3
  schedule of terms and prices, 109–11, 191
Authors, Society of, 10–12, 14–16, 18, 30, 31, 33, 39, 45, 46–8, 63, 67, 71, 80, 82, 84, 85, 96, 119–20, 123, 124–7, 129, 201–2

Bagnall, A., 211
Baker, John, 133, 135
Baldwin, Lord, 86
Barker, A. W. (Longmans), 123 n.
Barker, Sir Ernest, 182
Barker, R. E., 211, 211 n.
Barrington Ward, F. T., 71 n.
Baur, Karl, 131
Beith, Major I. H., 85
Bell, Edward (G. Bell & Sons), 25, 26 n., 28, 29, 32, 35, 38 n., 72
Bell & Sons, G., 60
Bell, C. F. Moberly, 23, 25, 27, 28, 30, 32, 34
Benn, Ernest, Ltd, 133
Bennett, Arnold, 62
Benson, A. C., 33

Bentley & Son, 7, 8
Bergne, Sir Henry, 32
Beveridge, Sir William, 97, 170
Bevin, Ernest, 183
Bibliographical practices, 12, 20, 42, 43, 101
Bickers, G. H. (G. Bell & Sons), 37 n., 90, 95 n., 96, 98, 99 n., 103, 154, 165 n., 171 n.
Binders, charges by, 67, 87, 156–8, 204
Binding
  boards and cloth, 57, 114, 162, 167, 174
  research, 204
Binyon, Sir Laurence, 31
Birchenough, Sir Henry, 55 n.
Birmingham City Corporation, 96
Black, A. & C., 7, 23
Blackmore, R. D., 106
Blackwell, Sir Basil, 136, 138, 142, 143, 143 n., 146, 150, 152 n.
Blackwood, J. H., 50, 52, 154 n.
Blackwood, William, & Sons, 7, 8
Book
  agents, 96, 152, 210
  clubs, 114, 134–8, 198
  jackets, 100, 177, 180
  Manufacturers' Association, 156–8
  Production War Economy Agreement, 171, 176, 185
  Society, 134
  Tokens, 114, 138–44, 197
  weeks, 62, 114, 153
Bookbinders' Association, Master, 67, 87
Bookseller, The, 62, 73, 89, 101, 105–8
Booksellers
  Association, 5, 7, 8, 12, 13, 15, 16, 21, 25, 43, 44, 45, 47, 59, 60, 71 n., 90, 94, 96, 98, 99, 102–8, 134, 135, 137–52, 154, 160, 192, 197–9, 210, 212
  decline of, 13 n.
  listing of, 100–4, 151, 152, 198, 208, 209
  London Committees, 1–4, 7
  Provident Institution, 92
Book-trade
  Directory, 101, 105, 108, 209
  periodicals, 101, 105–8
  surveys of, 99–105, 208
  distributive organization, 64–6, 98, 100, 187, 188, 200

222 Index

Boon, J. T. (Mills & Boon), 211 n., 212
Boot's Library, 45
Boriswood, 128
Bott, Alan, 134 n.
Bowes, G. B., 77, 96, 99 n., 104, 106
Bracken, Brendan, 175
British Broadcasting Company/Corporation, 84–7, 124, 125
British Council, 114, 117, 118, 175, 187, 188
British Federation of Book Publishers, 184
British Institute in Florence, 76
British Institute of Industrial Art, 74
British Italian League, 77
British Museum, 82, 83
British National Bibliography, 82 n.
British Publishers' Representatives in Australia, Association of, 111
Brown, Sir Cyril, 174
Brown, F., 99 n.
Brown, J. G. N. (Oxford University Press), 206, 211 n.
Brown, T., 174
Brunner, Elizabeth, 212
Buchan, John (Lord Tweedsmuir), 64, 123
Buckley, Mr Justice, 207, 211
Burghley, Lord (Marquess of Exeter), 167
Burn, James, & Co., 50
Butler, R. A. (Lord), 183
Buxton, Sydney, 39
Byard, T. (Heinemann), 94, 105, 107

Caine, Sir Hall, 31
Camara Oficial del Libro, 122
Cambridge University Press, 9 n., 64 n.
Campbell, Lord Justice, 1, 3, 4, 8, 211
Canada
  distributing centre, 196, 208
  import duties, 112 n., 115
  manufacture, 194
  market rights, 120, 190, 194–6, 208
  report on delegates' visit, 194, 195
Cannon, W. B. (Oxford University Press), 111 n., 167
Cape, Jonathan, 73 n., 99 n., 150 n., 165 n.
Cassell & Co., 89 n.
Catalogues, co-operative, 62, 76
Cercle de la Librairie, 17, 74, 78
Chadwyck-Healey, G. E., 65
Chamberlain, C. W. (Methuen & Co.), 99 n., 111 n., 146, 199 n.
Chamberlain, Joseph, 10
Chapman, John, 2, 3
China, copyright in, 80, 204
  Publishers Association, 204

Cholmondeley, Mary, 31
Christian, Bertram (Christophers; James Nisbet & Co.), 89, 115, 117 n., 126 n., 128 n., 150 n., 154 n., 166 n., 173, 181 n.
Churchill, Sir Winston, 25, 26, 57, 84
Circulating Libraries Association, 45
Clarke, James, & Co., 23
Clarke, W. H., 195
Clay, C. F. (Cambridge University Press), 55, 57 n., 82, 84 n., 110
Colleges, trade recognition of, 44, 96
Collins, William, Sons & Co., 21 n., 129, 165 n.
Collins, W. A. R., 165 n., 171 n., 195 n.
Colonial Office, 178
Commercial Lending Libraries Association, 199
Commons, House of, 3, 39, 123, 129, 168, 169, 183, 207
Constable & Co., 33, 146
Contemporary Review, 33
Cooper, Sir F. D'Arcy, 166 n.
Co-operative Societies, 97, 153, 198
Copenhagen Booksellers' Association, 65
Copyright
  agents, 11
  Association, 10, 18, 39
  Australian, 19, 112
  Canadian, 10, 19, 41, 60, 81
  Cases, 20
  during war-time, 52, 61, 202, 203
  imperial, 10, 11, 38, 39, 41, 42
  in abridgements, 10
  in extracts, 11, 41
  in mechanical reproductions, 39, 41, 84–6, 123, 125, 203
  in typography, 81, 123
  international convention, 10, 37, 38, 41, 42, 52, 60, 61, 80, 82, 121–3, 203
  Japanese, 121
  legislation, 1, 9, 10, 18, 19, 37–42, 61, 66
  libraries, 39, 40, 82–4
  litigation, 19
  registration, 19, 48, 82
  in translations, 41, 61
  U.S.A., 42, 60, 67, 80, 124, 203
  in unpublished work, 41, 82
Coupons, advertisement, 147–9; gift, 145–7
Cragg, C. W., 136
Crane, Edward M., 196, 197
Crossword puzzles, 89
Cumberlege, G. F. J. (Oxford University Press), 171 n., 195 n.
Cunningham, W., 15 n.
Curtis Brown, S., 188

Customs and Excise, 19, 48, 79, 114, 168

Daily Herald, 146
Daily Mail, 146
Dalton, Hugh, 176, 178, 183
Dane, Clemence, 134
Darling, Lord Justice, 33
Darwin, Sir George, 45
David, R. W. (Cambridge University Press), 210, 211 n.
Davies, R. G., 136
Davis, Elmer, 192
Dawson, William, and Sons, 76
de la Mare, R. (Faber), 162 n., 165 n., 171 n.
De Morgan, William, 45
Denny, F. A., 99 n.
Dent, F. J. Martin, 162 n.
Dent, Hugh, 133
Dent, J. M., & Sons, 21 n., 165 n.
Dickinson, C. H., 195
Dickson, Lovat (Macmillan), 195 n.
Dill, T. R. Colquhoun, 86 n.
Distributive organization, 63–6, 98–100
Dixon, W. MacNeile, 65
Documentary films, 153
Douglas, James, 29
Duckworth, G., 55, 84 n., 99 n.
Duncan, Sir Andrew, 184

Eady, Sir Wilfrid, 168
Eddington, Sir Arthur, 109, 169
Education Acts, of 1870, 4; of 1902, 20; of 1944, 177
Education Authorities, Local, 20, 21, 44, 49, 59, 60, 70, 90, 96, 102, 156, 177
Education, Board of, 56, 90, 156, 175
Education, Conference of Allied Ministers of, 178, 182, 202 n.
Educational
  books agreements, 22, 59, 90
  contractors, 20–2, 49, 59, 90
  Group, 70, 90, 91, 111, 155, 180, 205
Egypt, copyright in, 80
Elton, Lord, 184
Empire Press Union, 129
Employers' Federation, The Book Trade, 88, 89, 154, 155
Employment Circle, Book Publishers', 69, 88
  Group, 155, 205
Encyclopaedia Britannica, 23
Encyclopedias, copyright in, 19
English Catalogue, The, 101
Esher, Viscount, 33
Evans, C. S. (Heinemann), 37 n., 98, 99 n.
Evening News, 29

Export Council/Group, 166, 185, 205
Exports, book, 57 n., 75, 76, 115–20, 162, 176, 178, 184, 187–97, 207, 208
  American competition, 63, 118, 120, 121, 187, 189–97, 207
  to Australia and New Zealand, 109–12, 115, 163, 190, 191, 192, 193
  to Canada, 79, 115, 163, 185, 190, 194–6, 208
  to Commonwealth, 187, 189, 190, 196
  credit service, 116
  distributive organizations for, 63–6, 187–9
  licensing, 163
  research, 197
  to U.S.A., 79, 115, 194, 196
Eyre & Spottiswoode, 89 n.

Faber, Sir Geoffrey, 73 n., 117 n., 121 n., 125, 126 n., 128 n., 130, 135, 142, 159, 161, 164, 165, 168, 169, 171 n., 181 n., 193–5, 195 n., 199 n.
Fagan, B. W. (Edward Arnold & Co.), 165, 172 n., 181 n., 195 n., 196
Farquharson, W. (Murray), 37 n.
Fiction Group, 70, 71, 91, 109, 111, 119, 156, 199, 205, 210
Fine Art Guild, 66
Fisher, H. A. L., 56
Fisher, H. A. P., 211
Fishmongers' Company, 32
Florence Book Fairs, 76, 77
Foreign exchange, 75, 77, 115–17, 166 n., 191
Foreign Office, 63, 117
Forster, E. M., 128
Foyle, G., 99 n.
Francis, F. E. (Macmillan), 123 n.
Frere, A. S. (Heinemann), 165 n.
Freshfield, D., 32

Geddes, Sir Auckland, 58
Gentry, H. A. (Cassell), 195 n.
Germany
  book-trade organization, 51 n., 63, 98, 101
  Publishers Association, 116; Weimar resolution of, 131
Gill & Sons, 9 n.
Gladstone, W. E., 2, 3
Goffin, R. C. (Oxford University Press), 122 n.
Goldsmiths, Worshipful Company of, 130
Gollancz, Victor, 73 n., 133, 134, 136, 158, 173
Goodall, Thomas, & Co., 157
Gordon, G. S., 109, 134
Gorell, Lord, 129
Goschen, Lord, 31, 32

224 Index

Gosse, Sir Edmund, 45
Government, recognition by, 63, 64, 168, 175, 176
Graham, Sir Ronald, 77
Grant, D. A., 211
Greenham, R. G. Harvey, 90 n., 154, 155
Grierson, W., 65
Grote, George, 3
Guild Books, 181
Guillebaud, C. M., 5 n., 7

Hachette, Louis, 78, 79
Hadfield, John, 131 n., 201
Haggard, H. Rider, 10 n., 92
Hambleden, Lord, 168
Hanks, F. J., 99 n., 106
Hanley, James, 128
Hardy, Cozens, 11
Harrap & Co., 165 n.
Harrap, G. G., 84 n
Harrap, Walter G., 121, 136, 144, 147, 147 n., 163, 165, 167, 171 n., 175, 181 n., 188 n., 193-5
Heaton, Henniker, 30
Heinemann, William, 8, 13 n., 17, 34, 38, 42, 45 n., 46, 51, 53, 54, 55 n., 63 n., 72, 78
Herbert, Sir Alan, 128, 129, 169
Herschell, Lord, 10
Hewlett, Maurice, 45
Hill, Sir A. V., 169
Hitchcock, Curtice N., 191-3
Hockcliffe, M., 139 n.
Hodder & Stoughton 89 n., 165 n.
Hodder-Williams, R. P. (Hodder & Stoughton), 126 n., 150 n., 165 n., 195 n.
Hogg, D. M. (Viscount Hailsham), 71 n.
Holland, R. H. Code (Pitman), 172 n., 188 n., 195 n., 197 n., 211 n.
Home Office, 128
Hooper, H. E., 23, 25, 34
Howard, G. Wren (Jonathan Cape), 73 n., 91, 108, 115, 118, 121 n., 126, 128 n., 136, 143, 144, 151, 154 n., 161, 162 n., 167-9, 171 n., 173, 193-5, 197 n., 199, 199 n., 203
Hutchinson & Co., 89 n.
Hutchinson, G. T., 46
Hutchinson, Sir Herbert, 165
Hutchinson, Walter, 205
Hutton, R. S., 203

Iceland, 115
Imperial College, London, 44
Imports, book
  copyright in, 60, 61
  duty on, 66, 88, 113-15
  from Germany, 61, 77
  restriction of, in World War I, 60, 63; in World War II, 163, 191 n., 196
Indian Group, 122
market rights, 11, 18, 120
Information, Director of, 64
  Minister of, 175
Inge, W. R., 109
International Congress of Publishers, 10, 17, 20, 42, 78, 115, 122, 130, 131
International Directory of the Book Trade, 43
International Institute for Intellectual Co-operation, 76

Jackson, A. S., 136
Jackson, W. M., 23
Japan, 78, 121
Jeans, Sir James, 109
Jefferies, Richard, 92
Jerrold, D. (Eyre & Spottiswoode), 173, 195 n., 199 n.
John Bull's Other Island, 33
Joint Advisory Committee, The, 102, 149, 151-3, 175, 198, 209
Joint Book Trade Committee, The, 99-105, 149, 208
Jones, Kennedy, 34
Jonsson, S., 115,
Juvenile Group, 70, 71, 91, 153, 155

Keay, Henry W., 25, 32, 47
Kenyon, Sir Frederic, 82, 97, 104
Kingsford, R. J. L. (Cambridge University Press), 122 n., 165, 165 n., 166 n., 171 n., 175, 176, 179, 181 n., 182, 183, 197 n. 199, 199 n., 203, 206
Kipling, Rudyard, 109

Labour
  Ministry of, 170, 177, 183
  restrictions in World War I, 57; in World War II, 163, 170, 176, 177, 181-6, 207
Lane, Sir Allen, 133, 158
Lane, John (The Bodley Head), 27, 29, 36, 72, 73
Lang, Andrew, 92
Lang, Cosmo (Archbishop of Canterbury), 168
Laurie, T. Werner, 83
Laurier, Sir Wilfrid, 19
Law Officers of the Crown, 61, 83
Lee, Sir Sydney, 33
Leighton, D., 139
Leighton, Sir Robert, 171 n.

Index   225

Leipzig, centre of German book trade, 51 n., 63; Exhibition, 51
Lever, J., 211
Lewis & Lewis, 32
Libel legislation, 71, 128, 129
Libraries
    Public, 16, 43, 90, 95–7, 104, 105, 150–2, 210
    subscription/circulating, 22–35, 44, 45, 59, 151, 156, 198
    The London, 16
    Twopenny, 114, 149, 151, 198
Library Association, The, 22, 43, 96, 97, 104, 105, 152
Lindsay, Kenneth, 169
Listener, The, 86
Location of Retail Businesses Order, 175, 198
Lockyer, Sir Norman, 31
London Catalogue, 13
London County Council, 49
London Mercury, The, 74 n.
Longman, C. J., 7–13, 18, 28, 39, 46, 47, 65, 70, 72, 92
Longman, William (I), 2
Longman, William (II), 37 n., 63 n., 86, 99 n., 103, 104,¹ 121 n., 122 n., 146, 150 n., 151, 152 n., 154 n.
Longmans, Green & Co., 8, 28, 32, 117n., 160
Lords, House of, 38, 39, 129, 175, 183, 184
Lorna Doone, 25
Lowndes, Mrs Belloc, 45
Lucas, E. V. (Methuen), 31
Lynd, Sylvia, 134
Lyttelton, Oliver (Lord Chandos), 178

McDougall, A. J., 90; Educational Co., 21 n.
MacGillivray, E. J., 20, 37, 38, 41, 42 n., 60, 82, 84
Mackinnon, Lord Justice, 83
Maclehose, R., 13, 15
Macmillan & Co., 8, 60, 90 n., 165 n.
Macmillan, Daniel, 121 n., 122 n., 135, 165 n.
Macmillan, Sir Frederick, 5–7, 9, 13 n., 14, 18, 34, 35, 38, 39, 46, 47, 52–4, 56, 57 n., 72, 84, 92, 103, 111 n., 122, 145, 146, 150 n., 212
Macmillan, Harold, 91, 123, 154 n., 155, 165 n., 168
Mallon, J. J., 169
Manchester Municipal Technical School, 44
Mansbridge, Albert, 169
Map Group, 155, 210
Marshall, Alfred, 5–7, 212
Marshall, Sir Horace, 72
Marston, Edward (Sampson Low, Marston & Co.), 106
Marston, E. W., 106, 108

Marston, M., 98, 99 n., 139
Marston, R. B. (Sampson Low, Marston & Co.), 8, 9 n., 106
Maskew Miller, T., 109
Maynard, R. A. (Harrap), 171 n.
Medical Group, 70, 116, 155, 180, 205
Menzies & Co., 17
Meredith, W. M. (Constable), 52 n., 56, 58, 65, 94, 95 n., 103 n., 105, 107, 111 n.
Messagerie Italiane, 116
Metal supplies, in World War I, 57; in World War II, 177
Methuen, Sir Algernon, 46, 72
Metrication, 20
Milford, Sir Humphrey (Oxford University Press), 52 n., 55 n., 56, 63 n., 75, 86, 86 n., 87, 90, 121 n., 156, 165 n., 199, 199 n.
Mill, John Stuart, 3, 6
Milman, H. H., 3
Mining Association, The, 88
Moberly Pool, 172–4, 176, 178–9, 183, 185, 205
    Sir Walter, 173
Monkswell, Lord, 10
Monotype Corporation, 74
Moore, George, 83
Morison, Stanley, 74, 133 n., 171 n.
Morley, F. V. (Faber), 144
Morris, Sir Philip, 181
Morrison (Herbert), Lord, 166
Mudie's Library, 44, 156 n.
Mundanus Ltd, 133
Murray, John (III), 2
Murray, John (IV), 7–9, 12, 16, 17, 18, 27, 29 n., 33, 35, 38, 72, 92
Murray, Lieut.-Col. John (V), 72
Music Publishers' Association, The, 18, 38, 66, 80, 85

Nash, Ogden, 195
National Book Council/League, The, 73, 95, 98 n., 114, 138–44, 153, 169, 200
National Service Acts, in World War I, 57; in World War II, 177
National Union of Teachers, 49
Net Book Agreement, 16, 22–35, 43, 73, 96, 97, 101, 103–5, 109, 150–3, 200, 208–12
Newbolt, Sir Henry, 31, 65
Newdigate, B. H., 74 n.
Newspaper and Periodical Proprietors' Association, 65, 86
Newth, J. D. (A. & C. Black), 23 n.
New Zealand
    booksellers in, 49, 109–11
    Education Boards, 49

New Zealand *cont.*
  import licensing, 190
  market rights, 11, 63, 118
  sales tax, 115
Northcliffe, Lord, 34
Novels
  censorship of, 45
  prices of, 44-8, 71, 91, 109, 156

Obscene libel, 128
Ordnance Survey, 155
'Other traders', 47, 96, 103, 149, 152
Oxford University Press, 9 n., 165 n., 206

Page, Cadness, 139 n.
Page, Walter H., 72
Pall Mall Gazette, 14
Palmer, 2nd-Lieut. F., 72
Paper Control
  in World War I, 52-6
  in World War II, 161, 162, 165, 167, 171-85, 207
Pares, Richard, 165, 176
Parliamentary registers, 184
Patent Office, 61
Peake, Osbert, 176
Pearson, Sir C. Arthur, 34
Penguin Books, 46, 133, 158, 181
Philip, G., 155
Photographic reproduction, 81, 123
Pigott, A. E. (J. M. Dent & Sons), 122 n.
Piracy, in Australia, 19; in Far East, 118, 121; in India, 11, 19, 118, 122; in Near East, 19, 118, 122; in S. America, 121
Pitman, Sir Isaac, & Sons, 21 n., 35-6, 167
Pitman, Sir James, 167
Plant, Sir Arnold, 174
Postal Union rates, 20, 76
Poulten, W., 9, 18, 154
Power, A. D. (Pitman; W. H. Smith & Sons), 35, 50
Price Control Act, 175, 177
Prices
  level of, 4, 59, 60, 208
  net, 5-7, 13-16, 22-35, 59, 60, 70, 90, 91
  novels, 14, 59, 60, 209
  regulation of, 1-3, 5, 7, 8, 13-15, 22-35, 70, 209
Priestley, J. B., 128, 134, 169
Printers
  British Federation of Master, 67-9, 87, 112 n., 129, 171, 175, 185, 204
  production and warehousing charges, 68, 204
Printing and Allied Trades Research Association, 204-5
Production, Ministry of, 177, 179, 180, 181, 183

Prothero, Sir George, 31
Publishers' Advertising/Publicity Circle, 147
Publishers Association
  Annual Reports, 11 n.
  Constitution: Council, 9, 69, 74, 154; membership, 9, 18, 166, 177, 205; Rules, 8, 9, 69, 74, 154, 155, 205; Secretary, 9, 75, 155, 211 n.
  Groups, establishment of, 70, 90, 122, 154 n., 155
  jubilee, 206
  Members' Circular, 38 n., 70
  standing committees, establishment of: Agreements, 12; Australasian, 110; Colonial piracies, 11; Copyright, 38; Credit, 116; Educational Books, 21; Export, 63; General Purposes, 37; Title-pages, 12; Trade, 52
Publishers' Circle, 50, 52, 57, 70, 73, 74, 84, 94, 121
Publishers' Circular, The, 13, 62, 76, 89 n., 101, 105-8
Purchase tax, 67, 168

Quantity buying scheme, 152
Queen Victoria, The Letters of, 33, 35

Ranschburg, V., 79
Raymond, H. (Chatto & Windus), 37 n., 98, 99 n., 102, 107, 126 n., 128 n., 138, 139, 139 n., 144, 144 n.
Read, J. N., 99 n.
Reading, Marquess of, 32
Ready, A. W. (G. Bell & Sons), 205
Reith, Lord, 85, 86
Religious Group, 205
Religious Tract Society, 89 n.
Remaindering, 43, 102
  partial, 102, 149, 156, 199
Reprint series, 46-8, 59 n., 133
Restraint of Trade, Committee on, 151
Restrictive Practices Corut, 210-12
  Trade Practices Act, 209
Review copies, 96, 151
Review of Reviews, 11
Rights
  foreign, 11, 118-20, 127 n., 187, 189, 196, 197, 202
  imperial, 11, 63, 118, 120, 187, 190-7, 207
  reversion of, 12, 201, 202
  subsidiary, 12, 124-6
Ripon Hall Conference, 150
Rivington, C. R., 97 n.
Rivington, G. C., 99 n., 105, 111 n., 165 n.
Roberts, D. Kilham, 119 n., 201